# HYDROMETALLURGY

# HYDROMETALLURGY

## Theory

## Volume 1

MICHAEL NICOL
Murdoch University, Murdoch, WA, Australia

ELSEVIER

Elsevier
Radarweg 29, PO Box 211, 1000 AE Amsterdam, Netherlands
The Boulevard, Langford Lane, Kidlington, Oxford OX5 1GB, United Kingdom
50 Hampshire Street, 5th Floor, Cambridge, MA 02139, United States

ISBN: 978-0-323-99322-7

For information on all Elsevier publications visit our website
at https://www.elsevier.com/books-and-journals

*Publisher:* Candice Janco
*Acquisitions Editor:* Anita Koch
*Editorial Project Manager:* Mica Ella Ortega
*Production Project Manager:* Paul Prasad Chandramohan
*Cover Designer:* Miles Hitchen

Typeset by TNQ Technologies

Working together
to grow libraries in
developing countries

www.elsevier.com • www.bookaid.org

# Contents

## Student Resources

Please refer the below companion site for Typical solutions to the Case Studies at the end of chapters
https://www.elsevier.com/books-and-journals/book-companion/9780323993227

# Preface

Hydrometallurgy is the science and engineering of the recovery and refining of metals from ores, concentrates and other materials by a series of operations carried out in aqueous solutions.

It is one of the most inter-disciplinary subjects encompassing mineralogy, chemistry, physics, biology, chemical, metallurgical and materials engineering. This is what makes it such an exciting and interesting field in which to work and conduct research but also one of the most difficult for students to master. This observation is the main incentive for the conversion of the course notes for undergraduate, postgraduate and professional development courses delivered over many years into a suitable textbook that hopefully will assist in making the subject less onerous and more interesting.

As a practitioner and teacher of hydrometallurgy for most of my professional life, I never cease to be enthralled by the many new developments and ways of combining apparently individual operations into a single cohesive operation that takes, for example, gold ore containing less than a few grams of gold per tonne of ore and transforms this "dirt" into bars with a purity of greater than 99.9% with a recovery exceeding 90%.

One of the obstacles we all face when dealing with the industry is the wide range of units used from Troy ounces for precious metals to grams per tonne of $U_3O_8$, percent titanium as $TiO_2$ and parts per million in solution. In my teaching, I have always tried to mix the units to give students practice in dealing with a problem they will inevitably face in the world of extractive metallurgy. The reader will find a sprinkling of this philosophy throughout the book. Thus, concentration in solution is variously specified in terms of g $L^{-1}$, mg $L^{-1}$ (or ppm for dilute solutions), mol $L^{-1}$ or the shorthand version, M. My apologies to the purists who are yet to convert the real world to SI units.

My life as a hydrometallurgist has been both exciting and rewarding and I would do it all over again given the chance.

# Acknowledgements

For someone who prefers working in a lab, pilot-plant or operation to writing about it, this book did not come easily. I am indebted to the many under- and post-graduate students, academics and the professionals in research and the industry who have encouraged me to persist over many years. In particular, my gratitude to Gamini Senanayake who finally cajoled me into contacting the publishers from which there was no going back. He, together with my colleagues Nick Welham and Nimal Subasinghe, made significant contributions to several chapters in the book, and I wish to acknowledge their insightful input.

# Introduction

**Hydrometallurgical processes are concerned with the production of metals or metallic compounds from ores, concentrates or other intermediate materials by a sequence of operations carried out in aqueous solutions.**

Some metals and metallic products produced by processes which are fully or partially hydrometallurgical in nature are the following.
- Gold and silver
- Zinc and cadmium
- Copper, nickel, cobalt metals and sulphates
- Titanium dioxide
- Vanadium pentoxide and chemicals
- Alumina
- Ammonium Diuranate (ADU)
- Tungstic acid
- Salt
- Manganese dioxide
- Lithium chloride/carbonate
- Platinum group metals (Pt, Pd, Ir, Rh, Ru, Os)

> Identify a local producer and location of as many of the above products as possible.

## 1.1 Objective of a hydrometallurgical process

The ultimate objective of any extractive metallurgical process is to *economically* extract the valuable component of an ore or intermediate product. In particular, any viable hydrometallurgical process will have as its technical objective one or more of the following.
- To produce a metal directly from an ore, concentrate or pretreated concentrate
- To produce a pure metal or metal compound from a crude metal or metal compound

Hydrometallurgy. https://doi.org/10.1016/B978-0-323-99322-7.00002-8

- To minimize the use of reagents, water and energy in the production of metals
- To have minimal impact on the environment in terms of the disposal of solid and liquid wastes.

An overall extractive process consists of a number of interrelated steps not all of which are hydrometallurgical in nature. A simplified flow diagram for an overall process is shown in Fig. 1.1.

In order to meet these objectives, the inputs and outputs of any modern process have to be carefully considered and an example of some of the important overall characteristics are summarized in Table 1.1 for production of one tonne of zinc metal by the conventional roast-leach hydrometallurgical route (see Chapter 12) from a zinc sulphide concentrate in a plant producing $225000 \text{ t a}^{-1}$ of zinc.

Look up the chemical formula for hydronium jarosite.
What are some other jarosites?

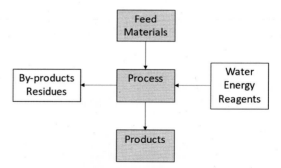

**Figure 1.1** An overall extractive metallurgical process.

## 1.2 Typical feed materials and products

One of the major advantages of hydrometallurgical processes is the ability to successfully recover valuable metals from a wide variety of feed materials such as.

### Low-grade ores

Typical gold ore containing between 1 and 5 g t$^{-1}$ gold.
Low grade copper ore containing 0.4% Cu with up to 10% Fe.
Lateritic nickel ore with 1.5% Ni, 0.2% Co, 70% $Fe_2O_3$
Low grade uranium ore with 200 ppm $U_3O_8$
Brines containing 0.5–1.5 g L$^{-1}$ lithium as chloride.

**Table 1.1 Important process parameters for zinc production.**

| Heat balance | MWh/t Zn |
|---|---|
| Steam production | 1.55 |
| Steam consumption | 0.88 |
| Losses | 19% |
| **Recoveries** | **%** |
| Zn | 92–98 |
| S as $SO_2$ | 92 |
| **Electrical energy consumption** | **MWh/t Zn** |
| Total | 3.8 |
| Electrowinning | 3.3 |
| **Residue to disposal** | **t/t Zn** |
| Jarosite[a] | 0.37 |
| Gypsum | 0.10 |
| Silicates, $PbSO_4$, $MnO_2$, etc. | 0.16 |
| S | 0.20 |

Note that the steam production comes from the energy liberated when the zinc sulphide is roasted (oxidised) in air to produce zinc oxide and $SO_2$ that is generally converted into sulphuric acid as a by-product. The high electrical energy consumption is due mainly to that used in the electrowinning of the zinc.
[a]Production of haematite or goethite would produce about 0.18 t gypsum/t zinc.

## Low-grade concentrates and calcines

Vanadium ore after magnetic concentration with 3% $V_2O_5$, 70% $Fe_2O_3$

## High-grade concentrates and calcines

Calcine from the roasting of a zinc concentrate containing 75% ZnO, 8% $Fe_2O_3$

Bauxite ore containing 50% $Al_2O_3$, 15% $Fe_2O_3$

## High-grade mattes

A matte (synthetic sulphide) produced by the smelting of a nickel sulphide concentrate containing 40% Ni, 30% Cu, 20% S, 1% precious metals.

## High-grade metals

Anodes for the production of copper by electrorefining containing 95% Cu, 2% Ni.

How much zinc (tonnes $a^{-1}$) could be produced in a plant treating 1000 t $d^{-1}$ of a concentrate containing 85% ZnS if the recovery in the plant is 95%?

The ores from which the valuable metals are extracted contain a large number of minerals which can be simple or very complex chemical compounds. Some of the most important of these are summarized in Table 1.2.

**Table 1.2  Important minerals treated by hydrometallurgical processes.**

| Metal | Mineral | Formula | % Metal by mass |
|-------|---------|---------|-----------------|
| Aluminium | Gibbsite, Boehmite | $Al_2O_3 \cdot nH_2O$ | Variable |
|  |  | $AlO(OH)$ | 62.8 |
| Cobalt | Erythite | $Co_3(AsO_4)_2 \cdot 8H_2O$ | 29.5 |
|  | Cobaltite | $CoS$ | 64.8 |
| Copper | Chalcopyrite | $CuFeS_2$ | 34.6 |
|  | Chalcocite | $Cu_2S$ | 80.0 |
|  | Enargite | $Cu_3AsS_4$ | 48.4 |
|  | Malachite | $Cu_2CO_3 \cdot (OH)_2$ | 57.5 |
|  | Chrysacolla | $(Cu,Al)2H_2Si_2O_5 (OH)_4 \cdot nH_2O$ | Variable |
| Gold | Native metal | $Au$ | 100 |
|  | Aurostibnite | $AuSb_2$ | 44.7 |
| Lithium | Spodumene | $LiAlSi_2O_6$ | 3.7 |
| Manganese | Hausmannite | $Mn_3O_4$ | 72.0 |
|  | Rhodochrosite | $MnCO_3$ | 47.8 |
| Nickel | Pentlandite | $(Ni,Fe)_9S_8$ | 34.2 |
|  | Millerite | $NiS$ | 64.7 |
|  | Gersdorffite | $NiAsS$ | 35.4 |
|  | Garnierite | $(Ni, Mg)_3Si_4O_{10}(OH)_2$ | Variable |
| Silver | Native metal | $Ag$ | 100 |
|  | Argentite | $Ag_2S$ | 87.1 |
| Uranium | Uraninite | $UO_2$ | 88.2 |
|  | Brannerite | $(U,Ce,Ca) (Fe,Ti)_2O_6$ | Variable |
| Titanium | Ilmenite | $FeTiO_3$ | 31.6 |
| Zinc | Sphalerite | $ZnS$ | 67.1 |
|  | Willemite | $Zn_2SiO_4$ | Calculate |

Calculate the % Zn in the mineral willemite.

The final products of a hydrometallurgical process can be either pure metals or metal compounds. Some examples are.

## Metals

Gold, copper, zinc, cadmium, nickel, cobalt, platinum, manganese and cadmium.

## Metal oxides

Alumina ($Al_2O_3$), manganese dioxide, vanadium pentoxide, nickel oxide.

## Metal salts

Cobalt or nickel sulphates, manganese sulphate, ammonium vanadate, ammonium diuranate, lithium carbonate.

In many cases, valuable by-products are also produced such as cadmium from the processing of zinc concentrates, gallium in the production of alumina, cobalt in the recovery of nickel and selenium, tellurium and the precious metals in the refining of blister copper produced by the smelting of copper concentrates, potassium carbonate in the production of lithium carbonate, and molybdenum in the recovery of uranium and copper.

The quality of the products produced by hydrometallurgical processes is continually increasing, and Table 1.3 summarizes the specifications for the chemical purity of two of the common products. Note the significantly higher purity required for copper due largely for the tight specifications required in the drawing of fine copper wire.

## 1.3 Hydrometallurgical process routes

Most extractive metallurgical processes can be divided into a series of so-called unit operations that involve the major divisions of mineral processing, hydrometallurgy and pyrometallurgy. Each process flowsheet used in a particular operation will depend on the size, nature and grade(s) of the ore deposit, the mineralogy of the ore, local infrastructure, availability of reagents and power

**Table 1.3 Specifications of high grade copper and zinc metals.**

| LME grade A copper | |
|---|---|
| **Element** | **Max.(ppm)** |
| Se | 2 |
| Te | 2 |
| Bi | 2 |
| Sb | 4 |
| As | 5 |
| Pb | 5 |
| S | 15 |
| Sn | 5 |

| Special high grade(SHG) zinc | |
|---|---|
| **Element** | **Max.(%)** |
| Al | 0.001% |
| Pb | 0.003% |
| Cu | 0.001% |
| Cd | 0.003% |
| Fe | 0.002% |
| Sn | 0.001% |
| Total | 0.005% max |

Note that 0.001% is equivalent to 10 ppm (parts per million).

and the nature of the products produced. Despite the fact that each plant is almost unique in the manner in which the various unit operations have been combined into an overall flowsheet, one can combine many of the operations into a single group of related operations. Thus, Fig. 1.2 shows a typical flowsheet for processes in which the ore is treated by hydrometallurgical operations, generally after crushing and milling and, in some cases, concentration by physical means such as screening, gravity, magnetic or flotation techniques. Typical examples of such processes would be the recovery of gold from ores, the so-called PAL (pressure acid leach) process for lateritic nickel ores, the recovery of copper from oxide ores by heap leaching, and the extraction of uranium and the production of alumina from bauxite ores. In some cases such as most gold operations, the ore is leached directly after crushing and grinding.

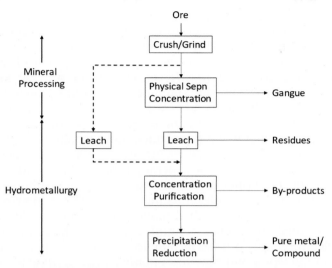

**Figure 1.2** A generic hydrometallurgical process for ores.

In many cases, the most profitable processing route can involve both mineral processing and pyrometallurgical operations prior to the use of hydrometallurgy as shown in the generic flowsheet in Fig. 1.3.

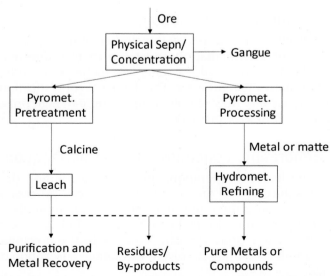

**Figure 1.3** Combined mineral processing, pyro- and hydrometallurgical processes.

In the left-hand option, an intermediate calcine product is produced by a high-temperature oxidation or reduction in the ore or a concentrate. This intermediate product is then more amenable to hydrometallurgical processing. Typical of such a route is the oxidative roast/leach/electrowin process for the extraction of zinc from zinc sulphide concentrates, the reductive roast/leach/precipitation process for the recovery of nickel and cobalt from lateritic ores by the Caron process and the thermal treatment of spodumene to enhance the leachability of lithium.

In the right-hand alternative, the pyrometallurgical process involves a smelting step that generally produces an impure metal such as blister copper or a matte (a mixture of synthetic metal sulphides) such as a nickel-copper matte. These intermediate products are subjected to hydrometallurgical processing to produce the final pure metals. Examples of this are the copper smelting/electrorefining process as practiced by many copper smelters around the world and the nickel matte smelting/pressure leach/reduction processes as practiced by the nickel and platinum producers.

## 1.4 Unit operations in hydrometallurgy

The major unit operations in hydrometallurgy can be subdivided into three main groups.

### 1.4.1 Leaching (or dissolution)

The selective dissolution of the desired mineral in an ore, concentrate or intermediate product can involve one or more of many different chemical reactants (leachants) and can be carried out in many different ways. The degree of sophistication and complexity can vary from simple heap leaching to high pressure/temperature autoclave processes.

### 1.4.2 Separation, concentration and purification

After the leaching operation, the resulting solution or pulp must be subjected to one or more chemical process steps designed to remove the impurities and/or concentrate the solution so that the desired metal can be successfully recovered in a pure form. These processes can involve selective precipitation, crystallization, cementation, solvent extraction, adsorption or ion exchange.

### 1.4.3 Precipitation and reduction

The final metal or metal compound is produced from the purified solution by a precipitation, crystallisation or reduction step depending on the desired product. The reduction step can involve electrons as in electrorefining or electrowinning but can also involve chemical reductants such as hydrogen gas.

Find out which leachants are used in the recovery of gold, uranium, vanadium and alumina. What are the final products of each of these hydrometallurgical processes?

The technology involved in the production of high-purity metals and materials from low-grade resources is continually evolving in order to satisfy the demands for greater recovery of metals of higher purity while satisfying the increasingly stringent environmental regulations for disposal of waste materials. Typical analyses for such materials in a conventional roast-leach-electrowin process for zinc are given in Table 1.4.

Production of metals of high purity requires that the solutions produced by leaching must be subjected to rigorous purification (and concentration in many cases) before the metal can be recovered by reduction. This is illustrated by the data in Table 1.5 which shows that, except for Mn, Mg and Ca the solution for the recovery of zinc by electrowinning is extremely pure.

**Table 1.4 Concentration range of elements in the major streams in the electrolytic zinc process.**

| Concentrate | | By-products | | Solid waste | | Liquid waste | |
|---|---|---|---|---|---|---|---|
| % | | % | | % | | % | |
| Zn | 45—60 | Ag | 0.001—0.05 | Fe | 2—14 | Mg | 0.02—0.3 |
| Cu | 0.1—2 | Pb | 0.2—3 | As | 0.01—0.4 | Cl | 0.01—0.03 |
| Cd | 0.1—0.4 | Hg | 0.001—2 | $SiO_2$ | 0.5—4 | F | 0.01—0.03 |
| S | 29—36 | Se | ppm | Al | 0.05—0.3 | $SO_4^{2-}$ | 0.1—1.0 |
| | | Ge | ppm | Mn | 0.1—0.6 | | |
| | | In | ppm | Ca | 0.05—1.5 | | |

**Table 1.5 Analysis of a typical purified leach solution for zinc electrowinning.**

| Metal | mg $L^{-1}$ | Metal | mg $L^{-1}$ |
|---|---|---|---|
| Zn(g $L^{-1}$) | 116 | Cd | 0.012 |
| Co | 0.03 | Tl | 0.025 |
| Ni | 0.08 | Mn(g $L^{-1}$) | 3.0 |
| Sb | 0.001 | Mg(g $L^{-1}$) | 10 |
| Ge | 0.001 | Ca | 54 |
| Cu | 0.07 | Fe | 3.5 |
| Cd | 1.0 | Al | 269 |
| As | <0.001 | Sn | 0.001 |
| Ga | 0.003 | Cl | 28 |
| Pb | <0.5 | F | 19 |

Environmental and occupational health and safety (OHS) requirements are placing important additional demands on the quality of the workplace and the quantity and nature of waste products produced by hydrometallurgical operations. Thus, the environmental problems associated with the disposal of residues containing toxic elements such as cadmium, mercury and lead have required that the zinc industry introduces new technology to produce residues that are stable in the environment. Table 1.6 summarizes some of the OHS issues in a typical zinc plant.

**Table 1.6 Some OHS issues in a typical zinc plant.**

| Process step/Source | Physical state | Main elements |
|---|---|---|
| Concentrate handling | Solid, dust | Zn, Fe, Cu, Cd, Hg, sulphides |
| Roasting | Solid, dust | Zn, Fe, Cu, Cd, Hg, oxides |
| Roasting | Gas | $SO_2$, $SO_3$ |
| Neutral leach | Solid, dust | Zn, Fe, Cu, Cd, Hg, oxides |
| Purification, Cd plant | Aerosol, liquid | Zn, Cu, Cd, sulphates |
| Waste water | Solution | Zn, Cu, Cd, Hg, sulphates |
| Electrowinning | Aerosol, liquid | $ZnSO_4$, $H_2SO_4$ |
| Melting/casting | Solid, dust | Zn, $ZnCl_2$, $NH_4Cl$ |

## 1.5 Description of a hydrometallurgical process

Hydrometallurgical processes generally involve the application of chemical reactions to the dissolution of solids, purification of the resulting solutions, recovery of pure metal products and the disposal of inert residues. In most of these operations, heterogeneous processes are involved and therefore transport of reactants to and products from interfaces that may be solid, liquid or gaseous is an integral part of the overall reaction.

A typical approach which is used in this text is shown schematically in Fig. 1.4. Thus, any chemical process requires a knowledge of the stoichiometric relationships between reactants and products without which it is not possible to assess the rates of metal production and reagent consumption and to undertake mass and energy balances. This requires some understanding of the nature of the chemical reactions taking place during the desired process but also the possible side reactions that inevitably accompany these processes. The chemical behaviour of the

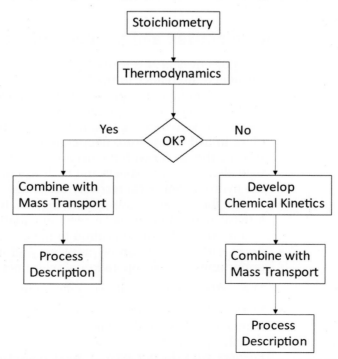

**Figure 1.4** Schematic procedure for the quantitative description of hydrometallurgical processes.

gangue (unwanted) components and of the trace metals can be major considerations in the development of a successful process.

The feasibility of a particular chemical system in achieving the desired outcome in a process step can be established by a suitable analysis of the thermodynamics (i.e. equilibria) of the system. This is a necessary but not sufficient condition for a viable process step and is generally carried out in the initial stages of development. A thermodynamic understanding can also often be of considerable value in defining suitable operating conditions for each process step.

In a limited number of operations, the kinetics of the chemical reactions taking place are sufficiently rapid that the rate determining step involves mass transfer of species to or from the reacting surface. Under these conditions, one would follow the left-hand column in Fig. 1.4 and a process description can be obtained by a combination of the chemical equilibria with mass transport.

On the other hand, the most frequent case will require an experimental study of the relevant rates of the chemical reactions which can then be combined with mass transport to produce a process description as shown in the right-hand column.

## 1.6 Objectives and structure of this textbook

The objectives of this textbook are to.
- Introduce the reader to the fundamentals of the science and the engineering of the processes used in hydrometallurgy.
- Enable the reader to make the necessary calculations to assess the thermodynamics and kinetics of the reactions that take place in hydrometallurgical operations.
- Provide the reader with the necessary tools to be able to evaluate alternative flowsheets and to devise novel approaches to the hydrometallurgical recovery of metals.
- Enable the reader to assess operating data from a hydrometallurgical plant and to make recommendations aimed at the optimization of or modification to the process.
- Assist with the solution of problems and implementation of opportunities in the operation of hydrometallurgical plants.

The textbook consists of two main components.

- Volume 1 contains Chapters 1–7 that deal with the fundamental chemistry and engineering which are required in order to fully appreciate and understand the subsequent materials. Readers who do not have a good background in chemistry and elementary chemical or metallurgical engineering would be advised to first complete these Chapters. Chapter 4 could be omitted by engineers.

- Volume 2 comprises Chapters 8–15 that deal with the main unit operations that comprise typical hydrometallurgical processes. This is the practical core of the course. With the exception of Chapters 4 and 8, each module is structured around the same major components of chemistry, thermodynamics, kinetics, engineering and examples.

Experience with the teaching of hydrometallurgy has revealed that the frequent use of problems based on the relevant theory can assist in the assimilation of the principles and application of the theory to real problems. There is therefore an emphasis on this aspect with both many problems and case studies at the end of each Chapter except Chapter 1. It is anticipated that worked solutions to the many problems and case studies will be posted on a dedicated website for readers of the text.

# 2

# Ions in solution

The chemical basis for hydrometallurgy is the science of ions in aqueous solution and in this chapter, we will summarise the chemistry of aqueous solutions with the focus on the nature of metallic ions in solution. Hydration (or aquation) of ions is the driving force for the dissolution of ionic solids and the structure of water as a solvent is important in our understanding of ions in solution. The objectives of this chapter are to provide the reader with an understanding of the important properties of metallic ions in aqueous solution as they relate to hydrometallurgical applications from an equilibrium point of view. We will do this by first summarising the complex interactions between water molecules and charged species. The properties and reactivity of metallic ions are governed by the structure of the aqua-ions, and this will be described in terms of inner- and outer-sphere co-ordination of water molecules. We will introduce the hydrolysis of metal ions from both a qualitative and quantitative perspective by reviewing the equilibria involved. The replacement of aqua-ions in the inner coordination sphere by other species (or ligands) to form various complexes will be introduced. The concept of hard and soft acids and bases will be used to qualitatively demonstrate the relative stabilities of these complex ions. The concept of useful species distribution diagrams will be developed, and the reader shown how these can be constructed with several examples from important hydrometallurgical processes. The problem of non-ideality of ionic solutions will be described and the activity coefficient as a conversion factor between ideal and non-ideal behaviour will be introduced. Various methods for the estimation of individual ionic activity coefficients will be summarised and pertinent examples provided.

## 2.1 Water as a solvent

Water is by far the most abundant liquid that occurs in nature. Although other solvents such as liquid ammonia possess very similar properties, water does dissolve an extensive number and type of substances. It is also stable in a particularly useful

Hydrometallurgy. https://doi.org/10.1016/B978-0-323-99322-7.00005-3

range of temperatures and pressures and its lack of objectionable properties and ease of handling make it an ideal medium in which to perform reactions and separations. Because of this and the many vital roles it plays in living systems, the properties of water have been extensively studied.

It is known that the two hydrogen atoms in a molecule of water are not bonded to the oxygen atom in a symmetrical arrangement with the bond angle being approximately 120 degree. Water thus behaves as a polar liquid with each molecule consisting of an electric dipole. This results in the formation of a relatively weak bond, the hydrogen bond, in which the hydrogen atom functions as a bridge between adjacent oxygen atoms. This results in the development of a partially ordered three-dimensional structure in which there are clusters of water molecules of various shapes and sizes depending on the temperature.

The introduction of a solid substance such as a sodium chloride results in sufficient electrical disturbance to rupture the hydrogen bond in the water at the interfacial surface. The released water dipoles are attracted to the points of charge on the solid and if these attractions are sufficiently large to overcome interionic attractions initially present in the solid, the substance will dissolve to form ions in solution. These ions are then surrounded (hydrated) by water molecules which form a fairly ordered structure close to the ion becoming less so as one moves to the bulk of the solvent. The resulting solution remains electrically neutral as the sum of charges on cations present is balanced by those on anions. Dissolution of an electrolyte such as hydrochloric acid in water results in strong hydration of the proton to form $H_3O^+$ and $H_9O_4^+$ ions in the so-called inner hydration sphere and weaker interactions as a result of hydrogen bonding in the outer hydration sphere as shown in Fig. 2.1. The chloride ion which forms only weak hydrogen bonds is weakly hydrated.

Fundamental theories on the properties of ions and other dissolved substances need to account for solvent—solvent, ion—solvent and ion—ion interactions at both short and medium range distances. Much experimental evidence has been collected and extensive theoretical work has been published in an attempt to predict the thermodynamic properties of aqueous solutions under conditions of practical interest in a field such as hydrometallurgy. Some consistency does now exist in theories that are limited to very dilute solutions, but even such a fundamental property as the electrical conductivity of mixed electrolytes with the total concentration of the order of a mole per litre is difficult to predict. This is because the above-mentioned interactions

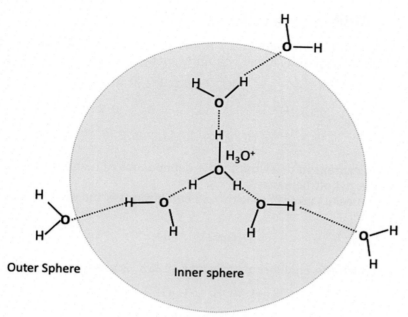

**Figure 2.1** Hydration of a proton.

become more complex in mixed electrolytes as the distance between adjacent ions and other species decreases. It is inevitable therefore that approximation will have to be made and a degree of empiricism introduced to obtain answers which are of practical interest.

The scientist or engineer who has to deal with problems, such as how a solution will behave when conditions of concentration, temperature, and pressure vary, cannot expect at this stage to obtain accuracies to better than about 20% in solutions of up to 3 molar concentrations. Computer packages are available which give detailed information about species present in solution and also enable mass and energy balances to be performed even when multiple reactions occur. These are critically dependent on the data bases used and how theories, both fundamental and empirical are applied to an actual system. It is of importance for users of such packages to be familiar with the limitations of the particular programs and to be able to do some order of magnitude calculations independently to see whether the results obtained are reasonable. We shall start with simple examples many of which are familiar to those with a chemical background.

## 2.2 Acid-base equilibria

Many of the reagents used in the processing of materials are acids or bases and simple equilibria such as

$$SO_4^- + H^+ = HSO_4^-, \quad logK_a = 1.91 \text{ at } 25°C$$

$$NH_3 + H^+ = NH_4^+, \quad logK_b = 9.24 \text{ at } 25°C$$

$$CN^- + H^+ = HCN(aq), \quad logK_b = 9.22 \text{ at } 25°C$$

are important in determining the dependence of many reactions on the pH of the system.

A special case of an acid-base equilibrium is the dissociation of water

$$H_2O \rightleftharpoons H^+ + OH^- \tag{2.1}$$

for which the equilibrium constant at 25°C is given by

$$K_w = (H^+)(OH^-)/(H_2O) = 10^{-13.99} \tag{2.2}$$

in which the terms in brackets are the activities (or concentrations in dilute solutions—see Section 2.11 for the distinction) of the species. $K_w$ is known as the ionic product of water and is often written as $pK_W = -log_{10}K_W$.

For a general case

$$A^- + H^+ = HA(aq) \tag{2.3}$$

one can write the equilibrium constant ($K_a$) for the formation of the acid and rearrange it to give a convenient relationship that enables one to locate the equilibrium on the pH scale.

$$(HA)/(A^-) = K_a \cdot (H^+) = 1 \text{ when } (H^+)$$
$$= 1/K_a (\text{or pH} = log_{10}K_a = -pK_a) \tag{2.4}$$

Sketch the curve for the ratio of the concentrations $HCN(aq)/CN^-$ as a function of pH.

## 2.3 Metal ions in solution

The behaviour of ions and their association with water molecules play an important role in the properties of electrolytes.

Although this interaction is not yet fully understood, particularly in concentrated solutions, it is clear that an ion such as $Ni^{2+}$ is surrounded by water molecules in a fairly ordered manner. Six molecules of water are attached by co-ordination bonds to the nickel atom in an octahedral configuration in what is known as the inner co-ordination sphere. Outside this sphere is a much larger number of more loosely held water molecules, which become less and less organised the further one moves towards the bulk solvent. Fig. 2.2 shows a two-dimensional representation of the inner sphere of the nickel ion in a solution of nickel sulphate.

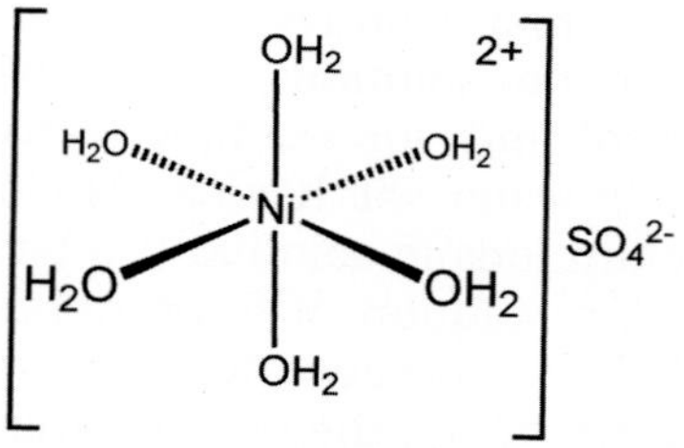

**Figure 2.2** Inner sphere co-ordination in the hexa-aquo nickel (II) ion.

In practice, we should write the formula for a nickel ion in solution as $Ni(H_2O)_6^{2+}$ but for convenience generally omit the water molecules.

The chemistry of ions in solution will depend largely on changes that affect the water molecules in the inner and outer spheres.

## 2.4 Hydrolysis of metal ions in solution

Taking the example of the hydrated nickel ion in Fig. 2.2, the first stage of hydrolysis occurs when a hydrogen ion (or proton) is removed from the inner coordination sphere

$$Ni(H_2O)_6^{2+} \rightleftharpoons Ni(H_2O)_5OH^+ + H^+ \qquad (2.5)$$

Ignoring the water molecules in the inner sphere we can write the reaction as

$$Ni^{2+} + H_2O \rightleftharpoons NiOH^+ + H^+ \qquad (2.6)$$

The second stage of hydrolysis follows in a similar fashion, that is:

$$NiOH^+ + H_2O \rightleftharpoons Ni(OH)_2 + H^+ \qquad (2.7)$$

The equilibrium involved in the first stage of hydrolysis of nickel can be expressed as

$$K_1 = (NiOH^+) \cdot (H^+)/(Ni^{2+}) \qquad (2.8)$$

in which (...) refers to the activity (or concentration for dilute solutions) of the species and $K_1$ is the first hydrolysis constant.

Taking logs to the base 10 and rearranging (remembering that $-\log_{10}(H^+) = pH$),

$$\log(NiOH^+)/(Ni^{2+}) = \log K_1 + pH \qquad (2.9)$$

that is, the ratio $(NiOH^+)/(Ni^{2+}) = 1$ when $pH = -\log K_1 = pK_1$

Table 2.1 gives the first hydrolysis constants of several important metal ions in dilute solutions at 25°C with calcium being the least 'hydrolysed' and, for example, a solution of calcium nitrate or chloride in water will have a pH close to neutral and almost all the calcium ions in solution will be as free $Ca^{2+}$ ions. On the other hand, a solution of ferric nitrate or chloride will have a pH well below 7 because ferric is strongly hydrolysed and, except at low pH values, the solution will contain very little free $Fe^{3+}$ ions.

What is the ratio of $(FeOH^{2+})/(Fe^{3+})$ at pH values of 1.2, 2.2, 3.2?

**Table 2.1 First hydrolysis constants at 25°C (Martell and Smith, 1976).**

| Metal ion | $-\mathrm{Log}_{10}K_1$ |
| --- | --- |
| $Ca^{2+}$ | 12.7 |
| $Mg^{2+}$ | 11.4 |
| $Mn^{2+}$ | 10.6 |
| $Fe^{2+}$ | 9.5 |
| $Co^{2+}$ | 9.6 |
| $Ni^{2+}$ | 9.4 |
| $Cu^{2+}$ | 7.5 |
| $Zn^{2+}$ | 9.1 |
| $Al^{3+}$ | 5.0 |
| $Fe^{3+}$ | 2.2 |

## 2.5 Formation of inner sphere complexes

Returning to our example of the nickel ion with its six water molecules in the inner co-ordination sphere, some or all of these molecules may be replaced by other groups or **ligands** (from the Latin 'ligo'-to bind). In the case of nickel chloride in ammonia solutions, Fig. 2.3 shows that all the water molecules have been replaced by $NH_3$ groups, but like hydrolysis, a sequence of complexes can be formed. Again omitting the water molecules

$$Ni^{2+} + NH_3 \rightleftharpoons Ni(NH_3)^{2+} \qquad (2.10)$$

$$Ni(NH_3)^{2+} + NH_3 \rightleftharpoons Ni(NH_3)_2^{2+} \qquad (2.11)$$

and so on to

$$Ni(NH_3)_5^{2+} + NH_3 \rightleftharpoons Ni(NH_3)_6^{2+} \qquad (2.12)$$

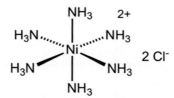

**Figure 2.3** The nickel hexammine $Ni(NH_3)_6^{2+}$ inner sphere complex.

In the ligand or coordinating species which may be anionic, neutral or even cationic, the atoms which are directly coordinated (attached) to the central metal ion are known as **donor** atoms as shown in Fig. 2.4 for nitrogen and oxygen. The metal ions that share the donated lone pair of electrons are known as **acceptor** atoms.

Thus, to give a few examples, nitrogen is the donor atom in ammonia, carbon or nitrogen in cyanide and oxygen in water. The nature of the metal to ligand bond, and hence the properties of the complex, are largely determined by the nature of the donor atom.

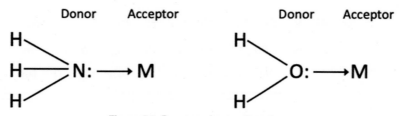

**Figure 2.4** Donor and acceptor atoms.

The following are some of the typical metal ions and non-metal ions that are encountered in hydrometallurgical processes.

- Simple hydrated ions

$$H^+, Li^+, Na^+, Pb^{2+}, Ni^{2+}, Cl^-, OH^-, SO_4^{2-}$$

- Weakly complexed ions or neutral species

$$CaSO_4, CuSO_4, CuCl^+, ZnSO_4, UO_2(SO_4)_2^{2-}$$

- Strongly complexed ions

$$PtCl_6^{2-}, Ni(NH_3)_4^{2+}, Ni(CN)_4^{2-}, Au(CN)_2^-, TiF_6^{2-}, CuCl_2^-, HSO_4^-$$

Metals and ligands are often classified as having so-called hard, soft or intermediate properties (Huheey et al., 2006). The position of elements on the Periodic Table gives an indication of this classification as shown in Fig. 2.5.

In terms of this qualitative theory, hard metal ions prefer bonding to ligands with hard donor atoms, and soft metal ions prefer soft donor atoms. Intermediate metal ions show no particular preference for either hard or soft donor atoms, forming

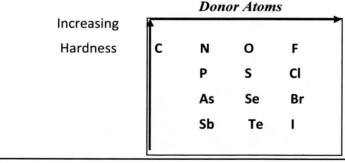

**Figure 2.5** Soft/hard donor atoms and metal ions.

reasonably stable complexes with both ligands. Hardness and softness are dependent on oxidation state in that it is usual for the ions of one metal to become harder as the oxidation state increases. Thus, for donor atoms, the hardest is fluorine while the softest is antimony. In the case of the metal ions, the proton is the hardest while the lead(II) ion is the softest.

There have been several attempts to quantify the degree of hardness or softness and the data in Fig. 2.6 shows selected data for both cations and anions from a compilation of the so-called hardness parameter, $H_B$, in the literature (Hancock and Martell, 1996). More positive values indicate increased degree of hardness.

Fig. 2.7 shows the effect of ligand softness on the $pK_{Sp}$ ($-logK_{Sp}$) (see Chapter 3 for definition of $K_{Sp}$) for various insoluble copper(I) compounds. The $Cu_2X_2$ or $Cu_2X$ compounds are more

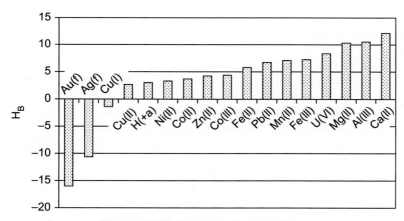

**Figure 2.6A** Hardness of selected cations.

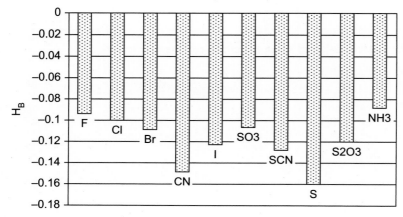

**Figure 2.6B** Hardness of selected bases.

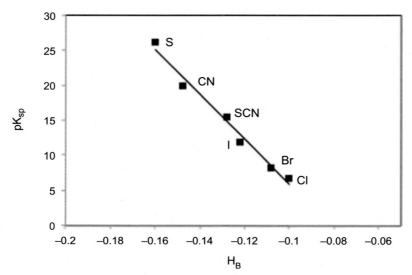

**Figure 2.7** Correlation between anion hardness and $pK_{SP}$ of insoluble Cu(I) compounds.

stable, that is have a lower solubility with anions ($X^-$) of greater softness such as sulfide ($Cu_2S$ $pK_{Sp} = 26$) than chloride ($Cu_2Cl_2$ $pK_{Sp} = 6$) because $Cu^+$ is a soft metal ion. These general free energy diagrams can be useful for the prediction of the thermodynamic properties where such data are not available.

Some particular examples of importance in hydrometallurgy are given in Table 2.2. One can see many examples of these effects. Thus, no oxides are known of the soft metal ion $Au^+$, a few sulphides are known, while gold telluride is stable enough to occur in nature. The refining of gold relies on the formation of a soluble complex $AuCl_4^-$ between the hard Au(III) cation and the hard chloride ligand. The effect of oxidation state is illustrated by $Cu^{2+}$ which is regarded as intermediate, while $Cu^+$ is soft.

**Table 2.2  Classification of some important cations and anions.**

|  | Acids | Bases |
|---|---|---|
| Hard | $H^+$, $Li^+$, $Na^+$<br>$Mg^{2+}$, $Ca^{2+}$, $UO_2^{2+}$<br>$Ti^{3+}$, $Cr^{3+}$, $Mn^{3+}$, $Fe^{3+}$, $Co^{3+}$, $Al^{3+}$<br>$U^{4+}$, $Si^{4+}$ | $NH_3$, $R\text{-}NH_2$<br>$H_2O$, $OH^-$, $O^{2-}$, $R\text{-}OH$, $-COO^-$<br>$CO_3^{2-}$, $NO_3^-$, $SO_4^{2-}$, $PO_4^{3-}$, $F^-$ |
| Borderline | $Fe^{2+}$, $Co^{2+}$, $Ni^{2+}$, $Cu^{2+}$, $Zn^{2+}$, $Pb^{2+}$ | $R\text{-}NH_2$, $NO_2^-$, $SO_3^{2-}$, $Cl^-$ |
| Soft | $Cu^+$, $Ag^+$, $Au^+$, $Cd^{2+}$, $Hg^+$, $Hg^{2+}$ | $CN^-$, $CO$, $S^{2-}$, $Te^{2-}$, $S_2O_3^{2-}$, $R\text{-}SH$ |

'Soft-soft' metal-ligand interactions are more favoured than 'soft-hard'; therefore, $Fe(NH_3)_2^{2+}$ is a known complex but $Fe(NH_3)_2^{3+}$ and $Al(NH_3)_2^{3+}$ are not known. Similarly, $Au(CN)_2^-$ is a very stable complex ion whereas the corresponding complex of cyanide with copper(II) is unstable. Similarly, 'hard-hard' metal-ligand interactions are more favoured than 'hard-soft', and therefore $AlF_6^{3-}$ is a stable complex ion but iron(II) forms only a very weak complex with fluoride.

Which ligands would be useful for the leaching of gold?

It must be emphasised that the above ideas of hardness and softness are merely a qualitative guide as to what possible inner sphere complexes could be formed. Quantitative information is obtained from formation (or stability) constants where these are available.

Thus, for the reaction (ignoring the water of hydration) of a metal ion $M^{z+}$ with a ligand $L^{y-}$ to form a complex ion $MX_n^{-(ny-z)}$

$$M^{z+} + nL^{y-} = ML_n^{-(ny-z)} \tag{2.13}$$

One can define the overall formation (or stability) constant

$$K = \beta_n = \left[ML_n^{-(ny-z)}\right] / \left\{[M^{z+}] \cdot [L^{y-}]^n\right\} \tag{2.14}$$

in which [....] denotes a concentration-generally mol/L.

In most cases, there are a number of equilibria involving the successive replacement of water molecules by the ligand in the coordination sphere of the metal ion

$$n = 1 : M^{z+} + L^{y-} = ML^{-(y-z)} \quad K_1 = \beta_1 = [ML^{-(y-z)}] / \{[M^{z+}] \cdot [L^{y-}]\} \tag{2.15}$$

$$n = 2 : ML^{-(y-z)} + L^{y-} = ML_2^{-(2y-z)}$$
$$K_2 = \left[ML^{-(2y-z)}\right] / \{[ML^{-(y-z)}] \cdot [L^{y-}]\} \tag{2.16}$$

and, for the overall reaction,

$$K_1 K_2 = \beta_2 = \left[ML^{-(2y-z)}\right] / \left\{[M^{z+}]g[L^{y-}]^2\right\} \tag{2.17}$$

and so on.

Note that $\beta_n = K_1.K_2...K_n$. that is $\beta_n$ is a cumulative stability constant and the greater the value of $\beta_n$, the greater the stability of the complex ion.

Examples of the formation of some common very strong, moderately strong and weak complexes are shown in Table 2.3. Appendix A8 contains a more extensive compilation.

Thus, the high stability of the aurocyanide ion is the basis for the cyanidation process for the extraction of gold while the formation of the hexamine complex of nickel(II) is important in several processes for the extraction of nickel. On the other hand, the weak sulphate complex of copper can often be ignored in the processing of this metal in sulphate solutions.

The acid-base properties of the metal and donor atoms are reflected in the values of the constants with hard-hard or soft-soft interactions giving rise to large constants. Note also that in the case of the nickel ammine complexes, the incremental $K$ values are approximately the same for the lower ammines. The values for the proton are important in that at low pH values, it will compete with the metal ions for the ligand and the extent

**Table 2.3 Selected stability constants (Martell and Smith, 1976).**

| Reaction | $Log_{10}K$ (at 298K) |
|---|---|
| $Ag^+ + 2Cl^- = AgCl_2^-$ | 5.25 |
| $Fe^{3+} + 2Cl^- = FeCl_2^+$ | 2.13 |
| $Au^+ + 2CN^- = Au(CN)_2^-$ | 38.8 |
| $Cu^+ + 4CN^- = Cu(CN)_4^{3-}$ | 23.1 |
| $H^+ + CN^- = HCN$ | 9.2 |
| $Al^{3+} + OH^- = Al(OH)^{2+}$ | 8.3 |
| $Al^{3+} + 4OH^- = Al(OH)_4^-$ | 34 |
| $Ti^{4+} + H_2O = TiO^{2+} + 2H^+$ | 12.5 |
| $Ni^{2+} + NH_3 = NiNH_3^{2+}$ | 2.72 |
| $Ni^{2+} + 2NH_3 = Ni(NH_3)_2^{2+}$ | 4.89 |
| $Ni^{2+} + 3NH_3 = Ni(NH_3)_3^{2+}$ | 6.55 |
| $Ni^{2+} + 4NH_3 = Ni(NH_3)_4^{2+}$ | 7.67 |
| $Ni^{2+} + 5NH_3 = Ni(NH_3)_5^{2+}$ | 8.34 |
| $Ni^{2+} + 6NH_3 = Ni(NH_3)_6^{2+}$ | 8.31 |
| $H^+ + NH_3 = NH_4^+$ | 9.24 |
| $Ag^+ + SO_4^{2-} = AgSO_4^-$ | 1.3 |
| $Cu^{2+} + SO_4^{2-} = CuSO_4$ | 2.36 |
| $Fe^{3+} + SO_4^{2-} = FeSO_4^+$ | 4.04 |
| $Fe^{3+} + 2SO_4^{2-} = Fe(SO_4)_2^-$ | 5.38 |
| $H^+ + SO_4^{2-} = HSO_4^-$ | 1.91 |
| $H^+ + OH^- = H_2O$ | 13.9 (pK$_w$) |

of complex formation will therefore be dependent on the pH in the case of many ligands such as sulphate, ammonia and cyanide. The last entry in the table is of particular importance as it defines the extent to which water is dissociated.

What is the fraction of silver present as $Ag^+$ in a 1M chloride solution containing 1 mM of total silver ions?

Some of the values of equilibrium constants for the formation of complexes of $U^{4+}$ and $UO_2^{2+}$ ions (both hard acids) with common ligands are listed in Table 2.4. It can be seen that the stability of complexes of $U^{4+}$ with $F^-$ is large due to favourable hard-hard acid-base interactions between $U^{4+}$ and $F^-$ ions, compared to the weaker hard-soft interaction between $U^{4+}$ and $NO_3^-$ ions. In the case of the common $UO_2^{2+}$ ion the stability of complexes follows the descending order $F^- > CO_3^{2-} > SO_4^{2-} > NO_3^-$ due to the decreasing hardness of the donor atoms in these anions. Several of these equilibria are important in the processing of uranium ores and spent nuclear fuel. Thus, the unusual stability of the uranyl carbonate complexes $\{UO_2(CO_3)_n^{-(2n-2)}\}$ is utilised in the leaching of uranium ores in carbonate/bicarbonate solutions.

Table 2.4 Equilibrium constants for reaction $U^{4+} + nX^- = U(X)_n^{-(n-1)}$ (Martell and Smith, 1976).

| n | $X = F^-$ (hard) | $X = NO_3^-$ (soft) |
|---|---|---|
| 1 | $10^7$ | $10^{-0.1}$ |
| 2 | $10^{13}$ | $10^{0.6}$ |
| 3 | $10^{17}$ | $10^{-0.4}$ |
| 4 | $10^{22}$ | $10^{-0.3}$ |

Equilibrium constants for reaction $UO_2^{2+} + nX^- = UO_2(X)_n^{-(n-2)}$

| Anion X | n=1 | n=2 | n=3 | n=4 |
|---|---|---|---|---|
| $F^-$ | $10^{4.5}$ | $10^8$ | $10^{10}$ | $10^{12}$ |
| $CO_3^{2-}$ | $10^{9.9}$ | $10^{16.6}$ | $10^{21.8}$ | — |
| $SO_4^{2-}$ | $10^{3.2}$ | $10^{4.1}$ | — | — |
| $NO_3^-$ | $10^{-0.4}$ to $10^{-0.7}$ | $10^{-1.7}$ | — | — |

The NTIS compilation that is the basis for the volume by Martell and Smith (1976) is a comprehensive tabulation of selected stability constants and has been used throughout this text. The constants tabulated in this compilation are generally based on concentrations in mol $L^{-1}$. Data for some complexes are given at various ionic strengths (see later). This is useful for more concentrated solutions. The compilation can be obtained at http://www.nist.gov/srd/nist46.

## 2.6 Formation of chelate complexes

There is a special class of complex ions known as chelate complexes in which a ligand (typically organic) is bonded to a central metal atom at two or more points. A simple example is the complex between a metal ion M and ethylene diammine shown in Fig. 2.8 in which two N-donor atoms are coordinated to the metal ion to form what is known as a bidentate complex.

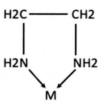

**Figure 2.8** Chelate complex of ethylene diamine with a metal ion M.

In the case of this ligand that has relatively soft N-donor atoms, it could be expected that strong complexes could be formed with soft metal ions such as $Cu^{2+}$. This is the case with log $\beta_1 = 10.6$. This can be compared with a value of log $\beta_2 = 6.6$ for the complex ion containing two unidentate ligands of methylamine ($CH_3NH_2$). Chelation therefore confers a greater stability on the complex than could be expected from the equivalent complex without chelation. A common example of this effect is haemoglobin in which four N-donor atoms from the same molecule are coordinated to $Fe^{2+.}$

The exceptional stability and selectivity of chelate complexes is utilised in the structure of reagents used to separate metal ions such as those of copper(II) and iron(III) in leach solutions produced in the heap leaching of copper ores (Chapter 11).

**Note:** In this text, we will write $Fe^{3+}$ for the ferric ion when we specifically refer to the aqua-ion. However, in most hydrometallurgical solutions, the predominant species is not the $Fe^{3+}$ ion but a complex ion with sulphate, chloride or other ligands. In these cases in which iron(III) is present as several different ionic or neutral species, we will refer to this species as iron(III) instead of $Fe^{3+}$ as has been done in the previous paragraph.

## 2.7 Formation of outer sphere complexes

Certain ligands are able to form complexes with metal ions in which they penetrate only as far as the outer co-ordination sphere, leaving the inner sphere untouched. The forces which hold a ligand such as the sulphate ion to the metal ion are mainly electrostatic with the aid of some hydrogen bonding. This complex formation, or ion-pairing as it is sometimes known, is generally weak but can have an influence on the behaviour of the electrolyte such as the activity of the various soluble species and the electrical conductivity of solutions.

More recent theories regard this association as a complex formed by the displacement or sharing of water molecules in hydration shells surrounding the two ions. Whichever mechanism is considered it is likely that the main force holding the ions together will be electrostatic in nature and that the formation constant for such an ion—pair or complex will be related to the charge product of the two ions.

The association between cations and anions of the general form $M^{n+}$ and $X^{m-}$ may be written as:

$$M^{z+} + X^{y-} \rightleftharpoons MX^{(z-y)+} \tag{2.18}$$

with equilibrium constant

$$K = [MX^{(z-y)+}]/\{[M^{z+}] \cdot [X^{y-}]\} \tag{2.19}$$

Consideration of the electrostatic interactions results in the following expression for K

$$\ln K = \ln K_o + \frac{e^2 yz}{r\varepsilon kT} \tag{2.20}$$

in which e = electronic charge.

k = Boltzmann constant

$\varepsilon$ = dielectric constant

r = distance between ionic centres.

$K_o$ = association constant of two uncharged species.

Estimate K for the interaction between $Ca^{2+}$ and $SO_4^{2-}$ ions.

The formation constants for a number of ion—pairs have been found to follow this relationship which suggests a linear relationship between ln K and the charge product 'yz' with Ln $K_0 = -1.8$ and slope $e^2/r \, \varepsilon \, kT = 1.8$. The effect of charge on the formation constant can be seen by the values for $Ag^+$, $Cu^{2+}$ and $Fe^{3+}$ with sulphate in Table 2.4.

## 2.8 Species distribution diagrams

As the name implies, these diagrams are very useful to summarise the various forms of a metal ion in solution. Consider the following equilibria involving the hydrolysis of the $Zn^{2+}$ aqueous ion at 25°C.

$$Zn^{2+} + OH^- = ZnOH^+ \quad \log K_1 (\text{or} \log \beta_1) = 5.0$$

$$Zn^{2+} + 2OH^- = Zn(OH)_2 \text{ (aq)} \quad \log K_1 K_2 (\text{or} \log \beta_2) = 8.3$$

$$Zn^{2+} + 3OH^- = Zn(OH)_3^- \quad \log K_1 K_2 K_3 \text{ (or} \log \beta_3) = 13.7$$

$$Zn^{2+} + 4OH^- = Zn(OH)_4^{2-} \quad \log K_1 K_2 K_3 K_4 \text{ (or} \log \beta_4) = 18.0$$

One can calculate the equilibrium concentrations of each of the five species as follows.
**(i)** Write the equilibria involved,

$$\beta_1 = [ZnOH^-]/[Zn^{2+}] \cdot [OH^-] = 10^{5.0} \qquad (2.21)$$

and similarly for the other three equilibria. Note that [....] denotes the concentration for dilute solutions.
**(ii)** Write the mass balance for zinc,

$$[Zn]_T = [Zn^{2+}] + [ZnOH^+] + [Zn(OH)_2] + [Zn(OH)_3^-] + [Zn(OH)_4^{2-}]$$
$$= [Zn^{2+}] + \beta_1 [Zn^{2+}][OH^-] + \beta_2 [Zn^{2+}][OH^-]^2 + \beta_3 \ldots\ldots + \beta_4 \ldots\ldots\ldots$$
$$(2.22)$$

where $[Zn]_T$ is the total zinc concentration. Re-arrangement gives,

$$[Zn^{2+}]/[Zn]_T = 1/\{1+ \beta_1[OH^-] + \beta_2[OH^-]^2 + \beta_3[OH^-]^3$$
$$+ \beta_4[OH^-]^4\} \qquad (2.23)$$

Thus, the fraction of zinc in the form of the $Zn^{2+}$ species can be calculated as a function of the pH (or $[OH^-]$).
**(iii)** Calculate the fractions of the other species as a function of pH using the above values of $[Zn^{2+}]$,

$$[ZnOH^-] = \beta_1 [Zn^{2+}][OH^-] \qquad 2.24$$

and similarly for the other species.

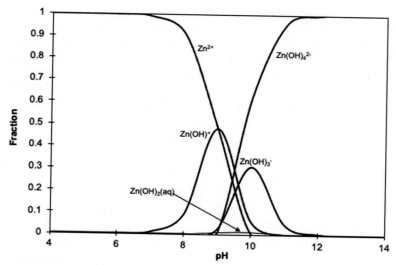

**Figure 2.9** Species distribution diagram for zinc in aqueous solution.

**(iv)** Plot the above fractions as a function of pH as shown in Fig. 2.9.

As could be expected, the relative proportion of the hydrolysed species increases as the pH increases while the simple aqua-ion is only stable at low pH values. This is generally true for all metal ions with the position of each species on the pH scale varying with the hardness of the metal ion.

We shall come across these diagrams again throughout this book. Note that the diagram can include more than one ligand although at the expense of computational complexity.

An important case in hydrometallurgy is that of sulphuric acid for which the following equilibria apply

$$SO_4^{2-} + H^+ = HSO_4^- \quad \log K_1 = 1.98$$

$$HSO_4^- + H^+ = H_2SO_4(aq) \quad \log K_2 \sim -3$$

For a specified acid concentration $[H_2SO_4]_T$, four unknown concentrations are required to construct a species distribution diagram. These concentrations can be obtained by solving a set of four equations, two of which define the above equilibria and two the mass balances for sulphate and the proton as follows

$$1. \ [HSO_4^-]/\{[SO_4^{2-}][H^+]\} = 10^{1.98} \qquad (2.25)$$

$$2. \ [H_2SO_4]/\{[HSO_4^-][H^+]\} = 10^{-3} \qquad (2.26)$$

$$3. \ [H_2SO_4]_T = [H_2SO_4] + [HSO_4^-] + [SO_4^{2-}] \qquad (2.27)$$

$$4. \ 2[H_2SO_4]_T = 2[H_2SO_4] + [HSO_4^-] + [H^+] \qquad (2.28)$$

These can be solved analytically but the expressions turn out to very unwieldy and difficult to use. A more convenient approach is to write the concentrations of all species in terms of say, $[HSO_4^-]$, $[H_2SO_4]_T$ and the constants. These can be entered in cells of an Excel spreadsheet together with a variable cell representing $[HSO_4^-]$. One can then use the Solver macro to solve for the value of $[HSO_4^-]$ that satisfies Eq. (2.27). The results are shown in Fig. 2.10 for various total acid concentrations.

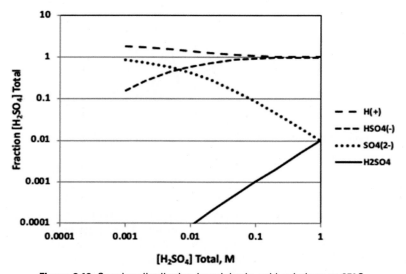

**Figure 2.10** Species distribution in sulphuric acid solutions at 25°C.

Except for very dilute solutions where it approaches twice the total acid concentration, the proton concentration is approximately the same as the total acid concentration. The same is true for the bisulphate ion concentration. The sulphate concentration is low at high acid concentrations and only approaches the total acid concentration at high dilution. The undissociated acid concentration is low over the whole range. Because of the low value for $K_2$, one can assume that $H_2SO_4$ is almost completely dissociated to bisulphate while bisulphate only dissociates to sulphate to a small extent in the practically important range of acid concentrations above about 0.1 M.

What would be the effect on the above distribution of adding sodium sulphate to a sulphuric acid solution ?

In Section 2.4, the donor atoms of the ligand (the molecule or ion that penetrates the inner hydration sphere of a metal) were classified according to hardness and softness. Three examples of such donor atoms will be chosen to demonstrate the behaviour of these complexes, namely C (soft), N (intermediate) and Cl (hard).

## 2.9 Some examples of species distribution

### 2.9.1 Cyanide

Cyanide complexes play an important role in the extraction of gold. In fact, if it were not for the introduction of the cyanidation process in 1890 to gold mines around the world, that industry would have had difficulty in expanding. It is probably the most successful hydrometallurgical process.

The reaction:

$$Au^+ + 2CN^- \rightleftharpoons Au(CN)_2^- \qquad (2.29)$$

has an equilibrium constant of log $\beta_2 = 38$, indicating that it is extremely stable complex. Most divalent base metal ions also form cyanide complexes, with the tetracyanide species generally being the most stable. Exceptions to these are $Fe^{2+}$, $Fe^{3+}$ and $Mn^{2+}$ which form complexes with six cyanide ligands. Monovalent metal ions show a tendency to form more than one complex as can be seen in Table 2.5.

**Table 2.5 Stability constants for some metal cyanide complexes (Martell and Smith, 1976).**

| Metal ion | Log $\beta_1$ | Log $\beta_2$ | Log $\beta_3$ | Log $\beta_4$ | Log $\beta_5$ | Log $\beta_6$ |
|---|---|---|---|---|---|---|
| Cu(I) | | 16 | 22 | 23 | | |
| Ag(I) | | 20 | 21 | | | |
| Au(I) | | 38 | | | | |
| Ni(II) | | | | | 30 | |
| Cu(II) | | | | | 25 | |
| Zn(II) | | 11 | 16 | 20 | | |
| Cd(II) | 6 | 11 | 16 | 18 | | |
| Fe(II) | | | | | | 35 |
| Fe(III) | | | | | | 44 |

The formation of relatively stable cyanide complexes with most base metal ions results in a significant consumption of cyanide (with the accompanying reagent cost) in a typical gold plant in the form of such complexes. In addition, the cyanide ion is not very stable and susceptible to other reactions particularly in ores containing reactive sulphide minerals. Fig. 2.11 shows such a deportment of various species of metal ion complexes and other decomposition products of cyanide in the tailings from a typical local gold plant. Note that the decomposition of cyanide is often catalysed by the metal ion complexes, notably those of copper.

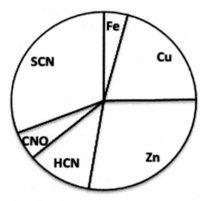

**Figure 2.11** A typical distribution of cyanide and its decomposition products during gold processing.

## 2.9.2 Ammonia

The ammonia ligand is fairly unique in hydrometallurgical systems of industrial interest in that the molecule is uncharged and therefore there are no significant electrostatic effects between it and the metal ion. The N donor atom will complex to varying degrees with a fairly large range of intermediate hardness metal ions.

Writing the general formation reaction as (divalent metal ions only):

$$M^{2+} + mNH_3 \rightleftharpoons M(NH_3)_m^{2+} \qquad (2.30)$$

with equilibrium constant

$$\beta_m = \frac{\left[M(NH_3)_m^{2+}\right]}{\left[M^2\right]\left[NH_3\right]^m} \qquad (2.31)$$

**Table 2.6 Stability constants of some metal amines (Martell and Smith, 1976).**

| Metal ion | Log $\beta_1$ | Log $\beta_2$ | Log $\beta_3$ | Log $\beta_4$ | Log $\beta_5$ | Log $\beta_6$ |
|-----------|------|------|------|------|------|------|
| $Ni^{2+}$ | 2.7 | 4.9 | 6.6 | 7.7 | 8.3 | 8.3 |
| $Co^{2+}$ | 2.1 | 3.7 | 4.8 | 5.5 | | |
| $Co^{3+}$ | | | | 25.0 | 30.1 | 34.4 |
| $Cu^+$ | 5.7 | 10.4 | | | | |
| $Cu^{2+}$ | 4.0 | 7.5 | 10.3 | 11.8 | | |
| $Zn^{2+}$ | 2.2 | 4.5 | 6.9 | 8.9 | | |
| $Mn^{2+}$ | 0.81 | | | | | |

Table 2.6 lists a typical set of values for the common base metal amines.

Although the value for manganese is small, the formation of an amine complex in solutions of high concentrations of ammonium sulphate is important in preventing precipitation of $Mn(OH)_2$ at the cathode during the electrowinning of manganese metal (see Chapter 14). Note the similarity of the constants for $Ni^{2+}$ and $Co^{2+}$. This similarity of behaviour of these two metals (this extends to the occurrence of their minerals in nature) makes separation of these two metals difficult. Note also the very much higher stability of the cobalt(III) amines. This results in rapid oxidation by atmospheric oxygen of cobalt(II) amine solutions to cobalt(III) that allows for an easier separation from nickel amines.

The species distribution diagram for nickel in ammoniacal solutions with a total ammonia+ammonium ion concentration of 1 M at 25°C is shown in Fig. 2.12 as fractions of the various species relative to the total nickel in solution as a function of pH.

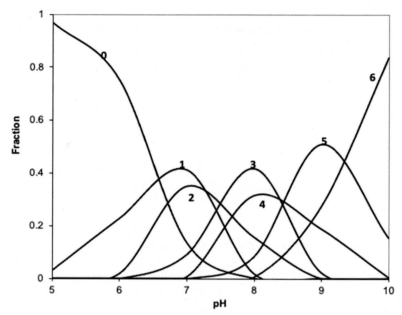

**Figure 2.12** Species distribution for the nickel amine complexes (Labels on each curve are values of m).

It is apparent that at pH 6, the dominant species is the aquo-ion $Ni^{2+}$ while at pH 7, the mono-amine is dominant with the higher ammines becoming more important as the pH increases.

The above diagram is important in defining the correct conditions for the processing of nickel matte(sulphide) in the Sherritt process which is practiced at a number of nickel refineries around the world. Thus, in the reduction of dissolved nickel amines to metal powder by the use of hydrogen gas under pressure, it is important to adjust the pH and the total ammonia concentration so that the diamine is the main nickel species in solution.

### 2.9.3 Chloride

Fig. 2.5 shows that fluoride is a ligand which shows a high degree of hardness. Thus, $Al^{3+}$ does not normally form strong complexes, but does so with $F^-$ for which log $\beta_6 = 21$ for the formation of $AlF_6^{3-}$. The formation of strong fluoride complexes of tantalum and niobium is used to good effect in the processing and separation of these metals.

On the other hand, chloride is somewhat softer and of particular importance are the chloride complexes of gold, silver and copper. Thus, the chloride complexes $AuCl_2^-$ (log $\beta_2 = 9.7$) and $AuCl_4^-$ (log $\beta_4 = 12$) are important in the refining of gold. The chloride route is followed for this process as silver is only sparingly soluble in a chloride medium and does not dissolve during electrorefining. The formation of relatively stable complexes of copper(I) with chloride ions but weak complexes of chloride with copper(II) results in copper(II) becoming a reasonably strong oxidant in chloride solutions as we shall see in Chapter 3.

Calculate the species distribution of copper(II) as a function of the concentration of chloride from 0.1 to 3M using the data in Appendix A9. It may be assumed that the concentrations of copper(II) are much smaller than that of chloride ions.

Solution:

The relevant chloride complexes and their constants are:

$$Cu^{2+} + Cl^- \rightleftharpoons CuCl^+ \quad \log \beta_1 = 0.4 \quad K_1 = 2.51$$

$$Cu^{2+} + 2Cl^- \rightleftharpoons CuCl_2 (aq) \quad \log \beta_2 = -0.4 \quad K_1 K_2 = 0.40$$

Writing the above equilibrium constants in terms of concentrations,

$$[CuCl^+]/[Cu^{2+}] = 2.51[Cl^-]$$

$$[CuCl_2]/[Cu^{2+}] = 0.4[Cl^-]^2$$

The mass balance for copper(II) requires that

$$[Cu(II)] = [Cu^{2+}] + [CuCl^+] + [CuCl_2]$$

$$= [Cu^{2+}] + 2.51[Cu^{2+}][Cl^-] + 0.4[Cu^{2+}][Cl^-]^2$$

from which,

$$[Cu^{2+}]/[Cu(II)] = 1/\left(1 + 2.51[Cl^-] + 0.4[Cl^-]^2\right)$$

These fractions of copper in the form of the uncomplexed ion can be obtained as a function of the chloride concentration and the corresponding fractions for the complexed ions obtained from the equilibrium constant expressions. The results are plotted in Fig. 2.13.

*Continued*

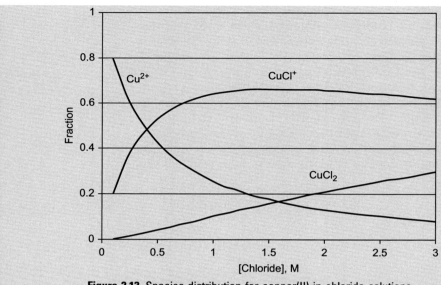

**Figure 2.13** Species distribution for copper(II) in chloride solutions.

It is apparent that at low chloride concentrations, most of the copper is present as the cationic aquaion but that at high chloride concentrations, the predominant species is the monochloro complex ion. Thus, extraction of copper(II) from chloride solutions by anionic extractants will be reduced at high chloride concentrations where the cation concentration is low.

## 2.10 Thermodynamics of ions in solutions

In designing hydrometallurgical processes, it is often necessary to perform energy balances and chemical equilibrium calculations. To determine these quantities, properties such as specific heat, enthalpy, entropy and free energy of formation will be required as functions of temperature, pressure and composition. This section is an introduction to the quantitative description of the thermodynamic fundamentals associated with ions in solution.

### 2.10.1 Enthalpy

The molar enthalpy of a species i at temperature T may be expressed as:

$$H_i(T) = H_i^o(T^o) + \int_{T^o}^{T} C_{pi}^o(t)dT + \Delta H_{vi} + \Delta H_i^l \tag{2.32}$$

in which $H_i^o(T^o)$ is the molar enthalpy of i at some specified basis condition, that is phase, temperature $T^o$ and pressure $P^o$. $C_{pi}^o$ (T) is the molar specific heat of i at the basis pressure $P^o$, $\Delta H_{vi}$ is the enthalpy associated with any change in phase and $\Delta H_i^l$ is a term representing the effect on the enthalpy of the change in pressure from $P^o$ to the operating pressure P.

For more than one component, the total enthalpy for a mixture will be:

$$H = \sum_i n_i H_i + \Delta H_m \qquad (2.33)$$

where $n_i$ is the number of moles of i and $\Delta H_m$ represents the effect on the total enthalpy of **mixing** the various components at temperature T and pressure P. The latter quantity can be significant in some cases such as the dilution of acids.

The basis condition or standard state is chosen such that the enthalpy of elements in their stable phase at $T^o = 298K$ and $P^o = 101.3$ kPa is zero. Since temperatures and pressures in hydrometallurgical processes have a limited range, it is reasonable to assume that mixtures of gases such as $O_2$, $N_2$ and $H_2$ are approximately ideal and that $\Delta H_i^l$ and $\Delta Hi_m$ may be taken as zero.

Mixtures that make up a solid phase may well be non-ideal. Examples are alloys of Ni and Cu, which may be electrowon from solution, as well as solid solutions such as $(Cu, Mn) CO_3$ which may form the feed to a leach circuit. Using the argument that pressure changes for solids and solutions are small, it is reasonable to assume that $\Delta H^l$ is zero and for most solid mixtures there is usually little error in taking $\Delta H_m$ as negligible. In specific cases where it is known that alloying and/or non-ideal solid solutions occur an attempt should be made to find data for these cases.

> Using the data in Appendix A2, estimate the enthalpy of the hydroxide ion at 90°C.

Aqueous solutions of ions pose a unique problem in thermodynamics. The reason being is that one cannot obtain a pure sample of an individual ion on which to do thermodynamic experiments. A new standard state has to be introduced to overcome this difficulty.

**The condition that has been chosen is that H and $C_p^o$ for $H^+$ ions are zero at all temperatures.**

Again $\Delta H^l$ may be taken as zero, but one could expect $\Delta H_m$ to be significant in concentrated solutions. From an overall energy balance point of view, it can be argued that in aqueous solutions which contain less than about 2 molar concentrations of salts, the enthalpy changes with respect to the water are significantly larger than the solutes and hence swamps most of the effects of $\Delta H_m$.

In order to calculate enthalpies of a particular mass of a mixture of substances or of a flow stream within a process, data for $H_i^o(T^o)$ and $C_p^o(T)$ are required as can be seen from Eqs. (2.32) and (2.33). In Appendix A data are given for a number of elements and compounds of interest in hydrometallurgy. More extensive data may be found in the references at the end of each Appendix.

Assuming that there is no phase change between $T^o$ and $T$ for any of the components and that $\Delta H^1 = 0$, the enthalpy in Eq. (2.33) reduces to:

$$H_i = H_i^o(T^o) + \int_{T^o}^{T} C_{pi}^o(T)dT \qquad (2.34)$$

For solids and liquids, heat capacity data is often presented in the form:

$$C_{pi}(T) = A_i + B_i T + C_i/T^2 \qquad (2.35)$$

where $A_i$, $B_i$ and $C_i$ are constants which may be obtained from tables (see Appendix A2). However, there are some components, particularly ions and neutral aqueous species where data are only available at 298K. Provided the temperature range is not too large, say $278K < T < 318K$, the enthalpy may be estimated from:

$$H_i = H_i^o(T^o) + (T\text{-}T^o)C_{pi}^o(T^o) \qquad (2.36)$$

in which $C_{pi}^o(T^o)$ is the specific heat of i at 298K and 101.3 kPa. Data for some of the more common ions and non-ionic aqueous species are also contained in Appendix A2.

## 2.10.2 Heats of mixing and dilution

So far we have been considering each substance to be a completely pure and separate material. The properties of an ideal mixture or solution can be calculated from those of the components. Thus, for a mixture of A, B,....

$$H = x_A \cdot H_A + x_B \cdot H_B + .... \qquad (2.37)$$

where $x_A$ is the mole fraction of A in the mixture.

When two or more substances are mixed to form an aqueous solution, we frequently find that heat is absorbed or evolved upon mixing. In this case,

$$H_{Soln} - H = H_{Mix} \qquad (2.38)$$

in which $H_{Soln}$ is the enthalpy of the final solution, $H$ is the above 'ideal' enthalpy calculated from that of the components and $H_{Mix}$ is the enthalpy of mixing. In the case of the solutions formed by the dissolution of one in another, this quantity is often called the enthalpy of solution. It has to be measured experimentally and has been tabulated for a number of common substances that we deal with in hydrometallurgy.

Thus, in Table 2.7, the heat of solution is expressed in terms of the enthalpy change per mole of solute (HCl in this case) as a result of consecutive additions of quantities of water to HCl.

**Table 2.7 Enthalpy of solution of HCl at 25°C and 1 atm (Sturtevant, 1940).**

| Composition | Moles H$_2$O/mole HCl | Incremental $-\Delta H$ (J/mol HCl) | Integral $-\Delta H$ (J/mol HCl) |
|---|---|---|---|
| HCl(g) | 0 | | |
| HCl 1H$_2$O(aq) | 1 | 26225 | 26225 |
| HCl 2H$_2$O(aq) | 2 | 22593 | 48818 |
| HCl 3H$_2$O(aq) | 3 | 8033 | 56851 |
| HCl 5H$_2$O(aq) | 5 | 7196 | 64047 |
| HCl 10H$_2$O(aq) | 10 | 5439 | 69486 |
| HCl 100H$_2$O(aq) | 100 | 4361 | 73847 |
| HCl 1000H$_2$O(aq) | 1000 | 837 | 74684 |
| HCl ∞ H$_2$O(aq) | ∞ | 213 | 75144 |

It is apparent that the heat evolved on adding a mole of water to an HCl solution decreases as the solution becomes more dilute. The integral enthalpy of solution is also shown and the value obtained at infinite dilution is called the standard integral enthalpy of solution. In the case of HCl and other acids, the relatively large enthalpy changes on dilution are due to the relatively strong energies involved in the hydration of the proton.

## 2.10.3 Free energies of aqueous ions

Formation of an ion from its elements can be represented as

$$Fe(s) = Fe^{2+}(aq) + 2e \qquad (2.39)$$

$$Mn(s) + 2O_2(g) + e = MnO_4^-(aq) \qquad (2.40)$$

Measurement of the accompanying energy changes is not possible and thermodynamic data is not available for the hydrated electron.

Therefore, the energetics of the formation of aqueous ions are defined relative to that of the reaction,

$$H_2(g) + 2e = 2H^+(aq) \qquad (2.41)$$

that is, for the above ions, $\Delta H$ and $\Delta G$ are the quantities associated with the reactions

$$Fe(s) + 2H^+(aq) = Fe^{2+}(aq) + H_2(g) \qquad (2.42)$$

$$Mn(s) + 2O_2(g) + 1/2H_2(g) = MnO_4^-(aq) + H^+(aq) \qquad (2.43)$$

This convention is equivalent to setting

$$\Delta H_f^o(H^+) = \Delta G_f^o(H^+) = 0 \qquad (2.44)$$

or

$$H_f^o(H^+) = G_f^o(H^+) = 0$$

Of course, this does not create any fundamental problem because we are always interested in the energetic **changes** accompanying chemical reactions and not the absolute values of the quantities.

For an isothermal reaction **with balanced charges** involving ions,

$$\Delta H^o = \Sigma \Delta H_f^o(\text{products}) - \Sigma \Delta H_f^o(\text{reactants}) \qquad (2.45)$$

$$\Delta G^o = \Sigma \Delta G_f^o(\text{products}) - \Sigma \Delta G_f^o(\text{reactants}) \qquad (2.46)$$

## 2.10.4 Entropies of aqueous ions

Although it is possible to measure the entropy change for a reaction such as,

$$NaCl(s) = Na^+(aq) + Cl^-(aq)$$

and the entropy of NaCl(s), it is not possible to assign entropies to each ion.

Once again, we introduce the convention that.

**The standard molar entropy of the proton (H$^+$(aq)) is zero at all temperatures.**

For an isothermal reaction **with balanced charges** involving ions,

$$\Delta S^o = \Sigma S^o(\text{products}) - \Sigma S^o(\text{reactants}) \qquad (2.47)$$

## 2.11 Activities of chemical species

Up to this point, we have used concentrations in all the equilibrium parameters. However, due to the many interactions between cations and other cations and anions in addition to the water molecules, metal ions do not behave as isolated species and the concentration does not necessarily reflect the actual 'activity' of the ion in solution, that is the ion does not behave ideally. Solution thermodynamics is designed to approximate reality in terms of deviations from ideal behaviour. The complex dependency of the activities on solution composition is thus dealt with by the introduction of a parameter known as the activity coefficient. Thermodynamic models are used to compute the activity coefficients of the solute species and the solvent.

For a solution containing ions, i, of molality (moles per kg solvent) $m_i$ and charge $z_i$, electroneutrality requires that

$$\Sigma m_i z_i = 0 \qquad (2.48)$$

with the summation carried out over all ions. The thermodynamic activities ($a_i$) of aqueous solute species are usually defined on the basis of molalities. Thus, they can be described by the product of their molal concentrations ($m_i$) and their molal activity coefficients ($\gamma_i$)

$$a_i = m_i \gamma_I \qquad (2.49)$$

The thermodynamic activity of the water ($a_w$) is always defined on a mole fraction basis. Thus, it can be described analogously by the product of the mole fraction of water ($x_w$) and its mole fraction activity coefficient ($\lambda_w$):

$$a_w = x_w \lambda_w \qquad (2.50)$$

The activity coefficients in reality are complex functions of the composition of the aqueous solution. In electrolyte solutions, the activity coefficients are influenced mainly by electrical interactions which are a function of the concentrations and electrical charges on the ions. Thus, a useful parameter is the ionic strength which is defined by:

$$I(or\ \mu) = 0.5\ \Sigma m_i z_i^2 \tag{2.51}$$

in which the summation is made over all aqueous solute species, $z_i$ is the electrical charge, and m is the molality(mol kg$^{-1}$ solvent). Except for concentrated solutions, the more convenient concentration C (mol L$^{-1}$ solution) can be used instead of the molality m. The two are related by

$$C_i = \frac{m_i \rho}{1 + M_i m_i} \tag{2.52}$$

in which $\rho$ is the solution density (kg L$^{-1}$) and $M_i$ is the molar mass (kg mol$^{-1}$) of the solute. For most dilute solutions in practice, $C_i \sim m_i$. For example a 1 molar aqueous solution of NaCl has a molality of 1.02 mol kg$^{-1}$ water.

Volatile aqueous species, which do not dissociate into ions (e.g., dissolved oxygen), may be assumed to obey Henry's law, particularly if their solubility is low.

In this case

$$a_i = k \cdot p_i / P \tag{2.53}$$

where k is a constant and $p_i$ the partial pressure of the species in the gas phase (total pressure P) in equilibrium with the solution.

## 2.11.1 Ionic activity coefficients

In electrolyte solutions, coulombic forces act between the ions due to electrostatic interactions over relatively long distances, that is ions do not behave as independent, isolated particles and this causes strong deviations from ideal behaviour even in relatively dilute solutions. At the equilibrium internuclear distance, the coulombic attractive forces between ions of opposite charge are balanced by strong repulsive forces between the overlapping electron clouds. Electrostatic forces also operate between molecules with permanent or induced dipoles and the charges on the ions. Thus, a polar solvent such as water can modify ionic behaviour and vice versa. As a result of the many complex interactions in aqueous solutions, no satisfactory theory of ionic solutions has been developed which will enable one to accurately predict the thermodynamic properties of solutions over wide ranges of concentration. Of course, for the same reasons given above about the thermodynamic quantities of individual ions, it is not possible to experimentally derive individual ion activity coefficients. As a result, various attempts have been made to theoretically (or empirically) calculate individual ionic activity coefficients.

Many models have been proposed for the estimation of $\gamma_i$ for ions in solution. These fall broadly into two categories, namely, theoretical and empirical. One of the first theoretical models to be derived was that of Debye and Huckel. Their expression for $\gamma_i$ is given by

$$\log\gamma_i = -0.5z_i^2\sqrt{I} \qquad (2.54)$$

This equation is valid only for very low concentrations and was superceded by the extended equation

$$\log_{10}\gamma_i = -\frac{Az_i^2\sqrt{I}}{1 + 1.5\sqrt{I}} \qquad 2.55$$

in which $A$ is a constant characteristic of the solvent (i.e. water) at the temperature and pressure. Values of $A$ for water at various temperatures and atmospheric pressure are given in Appendix A5.

Although the extended Debye-Huckel model was originally derived for dilute solutions ($I < 0.1$), some success has been reported in applying the extended version to higher concentrations, particularly when ion-pair equilibria are taken into account. Several modified versions of this equation have been used for more concentrated solutions. Some of the more common are discussed below.

## 2.11.2 The Davies equation

Although originally suggested for the calculation of mean ionic activity coefficients, this equation has been used to estimate single ion activity coefficients. It is a simple extended Debye-Hückel model in which an additional term proportional to I is included.

$$\log_{10}\gamma_i = -Az_i^2\left(\frac{\sqrt{I}}{1 + \sqrt{I}} - 0.3I\right) \qquad (2.56)$$

The coefficient 0.3 is sometimes replaced by 0.2 for more concentrated solutions. Note that it expresses the dependence on the solution composition through the ionic strength. The Davies equation is normally only used for temperatures close to 25°C. It is only accurate up to ionic strengths of a few tenths molal in most cases. The Davies equation has one great strength in that the only species-specific parameter required is the electrical charge. This equation may therefore readily be applied to a wide spectrum of species. Note that the Davies equation predicts a unit activity coefficient for all neutral solute species. This is known to be inaccurate. In general, the activity coefficients of neutral species that are non-polar (such as $O_2(aq)$, $H_2(aq)$, and $N_2(aq)$) increase with increasing ionic strength (the 'salting out effect',). On the other hand, the activity coefficients of some polar neutral species (e.g., the ion pairs $CaSO_4(aq)$ and $MgSO_4(aq)$) decrease with increasing ionic strength, presumably as a consequence of dipole-ion interactions.

### 2.11.3 The B-dot equation

A second model for activity coefficients is the B-dot equation of Helgeson for electrically charged species:

$$\log_{10} \gamma_i = -\frac{Az_i^2 \sqrt{I}}{1 + \dot{a}_i B \sqrt{I}} + (\dot{B}I) \tag{2.57}$$

in which $\dot{a}_i$ is the hard core diameter of the species, $A$, $B$ are the Debye-Hückel parameters, and $\dot{B}$ a characteristic B-dot parameter (0.04 at 25°C) that varies with temperature. Like the Davies equation, this is a simple extended Debye-Hückel model, the extension being the second term. The Debye-Hückel part of this equation is equivalent to that of the Davies equation if the product '$\dot{a}_i B$' has a value of unity. Values for $\dot{a}_i$, $A$ and $B$ are listed in Appendix A5.

The B-dot equation has about the same level of accuracy as the Davies equation. However, the B-dot equation can be used over a wide range of temperature (up to 300°C) which is useful in describing the thermodynamics involved in pressure leaching or reduction.

For electrically neutral solute species, the B-dot equation reduces to

$$\log_{10} \gamma_i = \dot{B}I \tag{2.58}$$

As $\dot{B}$ has positive values at all temperatures in the range of application, the equation predicts a salting out effect.

### 2.11.4 Pitzer's equations

Pitzer proposed a set of semi-empirical equations to describe activity coefficients in aqueous electrolytes. These equations have proven to be highly successful as a means of dealing with the thermodynamics of concentrated solutions containing uncomplexed metal ions. Pitzer's equations are based on a semi-theoretical interpretation of ionic interactions, and are written in terms of interaction coefficients (and parameters from which such coefficients are calculated). There are two main categories of such coefficients, 'primitive' ones which appear in the original theoretical equations, but most of which are only observable in certain combinations, and others which are 'observable' by virtue of corresponding to observable combinations of the primitive coefficients or by virtue of certain arbitrary conventions. Only the observable coefficients are reported in the literature.

There is a very extensive literature dealing with Pitzer's equations and their application in both interpretation of experimental data and in modelling. A complete review is beyond the scope of this text.

## 2.11.5 Meissner's method

Meissner developed a method based on a number of equations in which a 'reduced' activity coefficient ($\Gamma = \gamma^{1/z+z^-}$) is related to the ionic strength.

The relationship between $\Gamma$ and I was based on an observation that at 25°C values of the former for typical strong electrolytes in aqueous solution vary with the latter for ionic strengths up to 20. The curves fall on to a family of curves characteristic of the type of electrolyte. Thus, if $\Gamma$ is known for a particular ionic strength, at a given concentration, the entire curve can be estimated.

The reader is referred to the relevant reference given at the end of this Chapter for full details.

## 2.11.6 Solutions of non-electrolytes

In general, it is necessary to consider the case of solutions containing non-electrolyte solute species in addition to ionic species. Examples of such uncharged species include molecular species such as $O_2(aq)$, $CO_2(aq)$, $H_2S(aq)$, and $SiO_2(aq)$; strongly bound complexes, such as $HgCl_3(aq)$ and $UO_2CO_3(aq)$; and weakly bound ion pairs such as $CaCO_3(aq)$ and $CaSO_4(aq)$. The theoretical treatment of these kinds of uncharged species is basically the same.

In many cases such as dissolved gases, one can use a simple relationship such as,

$$\gamma = 0.15I$$

There are practical differences, however, in fitting the models to experimental data. This is simplest for the case of molecular neutral species. In the case of complexes or ion pairs, the models are complicated by the addition of corresponding mass action equations.

### 2.11.7 Individual ionic activity coefficients: pH scales

As we have seen in previous sections, it is not possible to measure any of the thermodynamic functions of single ions, because any real solution must be electrically balanced. Thus, the activity coefficients of aqueous ions can only be measured in electrically neutral combinations. These are usually expressed as the mean activity coefficients of neutral electrolytes. The mean activity coefficient of an electrolyte is given by

$$\gamma_{\pm} = \left[\gamma_+^{z+}\gamma_-^{z-}\right]^{1/(z^+ + z^-)} \tag{2.59}$$

where $\gamma_+$ and $\gamma_-$ are the individual cation and anion activity coefficients, which for a 1:1 electrolyte gives, $\gamma_{\pm} = (\gamma_+\gamma_-)^{1/2}$. It is often assumed that $\gamma_+ = \gamma$, which enables individual ionic activity coefficients to be derived from tabulations of mean ionic activity coefficients. However, this is generally not true because the more strongly hydrated cations are subject to greater variations in activity than the weakly hydrated anions.

Individual ionic activity coefficients can be defined on a conventional basis by introducing some arbitrary choice. This can be made by adopting some expression for the activity coefficient of a single ion. The activity coefficients of all other ions then follow via electroneutrality relations. Because this applies equally to the hydrogen ion, such an arbitrary choice then determines the pH. Such conventions are usually made precisely for this purpose, and they are generally known as pH scales. The NBS pH scale, which is the basis of nearly all modern conventional pH measurement, is based on the activity coefficient of the chloride ion using a modified Debye-Huckel equation

$$\log_{10}\gamma_i = -\frac{Az_i^2\sqrt{I}}{1 + 1.5\sqrt{I}} \tag{2.60}$$

An alternative approach involves the so-called Bates equation which for a 1:1 electrolyte such as NaCl can be used to estimate the single ion activity coefficients as follows,

$$\log \gamma + = \log \gamma \pm -h/2.\log a_{H_2O} \qquad (2.61)$$

$$\log \gamma + = \log \gamma \pm +h/2.\log a_{H_2O} \qquad (2.62)$$

in which $a_{H2O}$ is the activity of water and 'h' is the hydration number of the salt (cation plus anion).

Fig. 2.14 illustrates the variation of the mean ionic activity coefficients with increasing concentration for several electrolytes at 25°C. Note the effects of concentration and charge. The lower activity coefficients of the sulphate species are due to the higher ionic strengths compared to 1:1 electrolytes and also to some ion-association or ion pairing. In contrast, the chlorides have higher activity coefficients, which increase at high concentrations as a result of the decreasing water activity. In the case of HCl, the activity coefficient in 10 molal HCl is 10.5. This high value is the reason for the very high reactivity of concentrated HCl solutions.

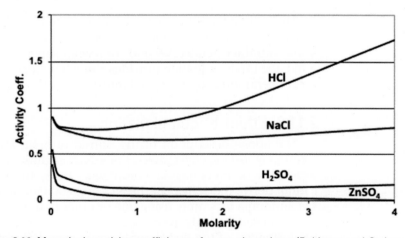

**Figure 2.14** Mean ionic activity coefficients of some electrolytes (Robinson and Stokes, 1959).

The data in Fig. 2.15 shows a comparison of the measured mean ionic activity of HCl as a function of ionic strength (concentration is the same in this case) with values estimated using several of the methods described above.

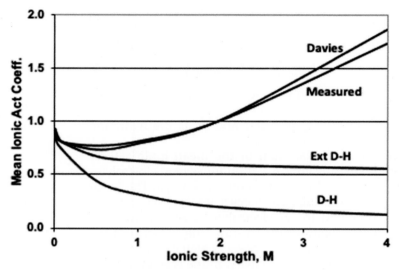

**Figure 2.15** Comparison of actual and calculated mean ionic activity coefficients of HCl at 25°C. D-H is Debye-Huckel.

The inadequate fit of actual and calculated coefficients for the two Debye-Huckel equations is obvious at millimolar concentrations with the extended version somewhat superior. On the other hand, the Davies equation provides a much better correspondence for concentrations up to several mol/L.

## 2.11.8 Effect of complexation

The above description of the non-ideal behaviour of ions in solution ignores specific interactions such as those involved in the formation of metal-ligand complexes. Thus, for example, one cannot derive the activity of the $Ni^{2+}$ aqueous ion in a solution of 1M ammonia by simply using one or other of the above methods. The results will greatly overestimate the activity of the ion because the formation of the nickel amine complexes has not been taken into account. One way around this problem is to use the following iterative method.

1. Use the method described above to calculate the concentrations of the various nickel species as shown in Fig 2.12 assuming initially that all activity coefficients are unity.
2. The ionic strength of the solution can then be calculated taking into account all of the ionic species present.
3. The ionic strength can in turn be used to estimate the activity coefficients of each species using, say, the extended Debye-

Huckel equation. In some cases, one will have to make a guess-timate for the ionic parameter $a_i$ by analogy with similar ions that are listed in Appendix A6.

4. These estimates can then be used to calculate new activities for each of the species that can be used in a second species distri-bution calculation. This will result in a modified distribution.

5. Use the modified distribution to repeat steps 2 to 4 and repeat until convergence is attained.

In some cases, it is possible to avoid this rather convoluted process if stability constant data for a complexation reaction is known at various ionic strengths. Thus, the NIST compilation referred to at the end of Section 2.5 does, in many cases, provide such information at various ionic strengths and these values (interpolated if necessary) can be used in place of calculated activity coefficients. Note that many of the quoted experimental stability constants are shown as being extrapolated to zero ionic strength, that is infinitely dilute solutions where, of course, the activity coefficients are unity.

The measured pH of 2 mol $L^{-1}$ HCl at 298K is $-0.63$. Calculate $a_H^+$, $\gamma_H^+$ and $a_{OH^-}$ of the solution.

## 2.12 Thermodynamic properties of ions at high temperatures

The properties of ions such as stability constants and activity coefficients are dependent on the temperature that we have assumed to be 25°C up to this point. The starting point in the estimation of these properties at elevated temperatures often used in hydrometallurgical processes is the Gibb's relationship,

$$\Delta G^o = \Delta H^o - T\Delta S^o \qquad (2.63)$$

that is suitable for limited temperature ranges($<50°C$), because $\Delta S^o$ and, particularly, $\Delta H^o$ are dependent on temperature,

$$\text{i.e. } \Delta H_T^o = \Delta\Delta H_o^{298} + \int_{298}^{T} Cp\,dt \qquad (2.64)$$

and similarly for $\Delta S_T^o$.

Heat capacity data are often not available in a form which makes the integration possible, so an average value $\Delta C_p^o]^T_{298}$ is often used or, if $C_p$ is only available at 298K, $\Delta C_p^o$ at 298K is used instead of the mean value.

Substitution in the above gives,

$$\Delta G_T^o = \Delta H_{298}^o - T\Delta S_{298}^o + (T\text{-}298)\Delta C_p^o\Big]^T_{298} - T\Delta C_p^o\Big]^T_{298} \ln(T/298)$$

$$(2.65)$$

The Gibb's equation predicts a linear increase of $\Delta G^o$ with temperature while the form is generally non-linear over extended temperature ranges. These complex thermodynamic relationships are often taken into account in software programs to calculate the effect of temperature on equilibrium constants.

Fig. 2.16 shows the effect of temperature on $pK_w$ and pH of water. Both $pK_w$ and pH decrease with increasing temperature, but the change in pH is relatively small at temperatures above about 100°C.

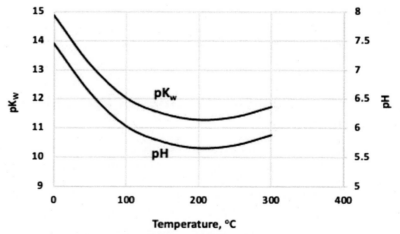

**Figure 2.16** Effect of temperature on $pK_w$ and pH of water.

Determine the equilibrium constant for the dissociation of water at 75°C using the above Eq. (2.65) and the data in Appendix A4.

The development of industrial processes such as high temperature pressure leaching of sulphide and oxide ores and mattes and the hydrogen reduction of ammoniacal nickel and cobalt solutions requires accurate thermodynamic data for ions at elevated temperatures. However, such data are not available except for a few common species over an extended temperature range. A method for predicting such data was devised by Criss and Cobble that is known as the 'Correspondence Principle'. This will be dealt with in Chapter 3.

## 2.13 Summary

This Chapter has introduced the important topic of the nature and behaviour of ions in solution as the basis for hydrometallurgical operations.

The reader should now have a fundamental understanding of some of the important properties of ions with the focus on metallic ions. The hydration of ions varies with the nature and charge on the ions. Substitution of water molecules by other ligands in the coordination sphere of ions results in ions that can have very different properties that can be exploited in the design and control of processes such as dissolution and separation. The qualitative theory of hard and soft acids and bases has been introduced as a method to explain and predict the relative stabilities of metal ion complexes. The reader should now be able to recognise and possibly exploit the relative interactions between metallic ions as acids and ligands as bases. The quantitative description in terms of stability constants is essential in the further development of such interactions and the reader should be able to manipulate and use published stability constant data to, among others, construct and interpret species distribution diagrams. These have been illustrated with several examples pertinent to common hydrometallurgical processes.

The lack of ideal behaviour of ions in solution at the concentrations appropriate to hydrometallurgical processes can be overcome by the introduction of the activity coefficient and various methods have been introduced that will enable the reader to estimate activity coefficients for individual ions even in relatively concentrated solutions at elevated temperatures encountered in practice.

# References

Bamer, H.E., Scheuennan, R.V., 1978. Handbook of Thermochemical Data for Compounds and Aqueous Species. Wiley–Interscience, New York.

Barin, I., Knacke, O., 1973. Thermochemical Properties of Inorganic Substances. Springer–Verlag, Berlin.

Criss, C.M., Cobble, J.W., 1964. Thermodynamic properties of high temperature aqueous solutions. J. Am. Chem. Soc. 86, 5385–5393.

Hancock, R.D., Martell, A.E., 1996. J. Chem. Educ. 73, 654–661.

Huheey, J.E., Keiter, E.A., Keiter, R./L., Medhi, O.K., 2006. Inorganic Chemistry: Principles of Structure and Reactivity, fourth ed. Pearson, New Delhi.

Kreith, F., 1976. Principles of Heat Transfer, third ed. Intext Press, New York.

Martell, A.E., Smith, R.M., 1976. Critical Stability Constants, vol. 4. Springer, New York.

Naumov, G.B., Ryzhenko, B.N., Chodakovsky, I., Barnes, I., 1974. Handbook of Thermodynamic Data. National Technical Information Service, U. S. Department of Commerce, Springfield.

Robinson, R.A., Stokes, R.H., 1959. Electrolyte Solutions, second ed. Butterworths, London.

Sturtevant, J.M., 1940. J. Am. Chem. Soc. 62, 584–587.

Zemaitis, J.F., Clark, D.M., Rafal, M., Scrivner, N.C., 1986. Handbook of Aqueous Electrolyte Thermodynamics. American Institute of Chemical Engineers, New York.

# Further reading

Kusik, C.L., Meissner, H.P., 1975. Calculating activity coefficients in hydrometallurgy–A review. Int. J. Miner. Process. 2, 105–115.

# Practice problems

The following problems are not complex and are here to provide a refresher of your chemistry. Also see Chapter 3 for a number of more relevant practice problems based on the material in Chapters 2 and 3.

1. Calculate the concentration of $H^+$ and $OH^-$ in a solution of pH = 3.5 at 298K, assuming unit activity coefficients, $pK_w = -Log\ K_w = 14$.

2. Hydrogen sulfide dissolved in water, is in equilibrium with $H^+$, $HS^+$ and $S^{2-}$ ions according to the following equilibria:

$$H_2S(aq) = H^+ + HS^-; \rightarrow pK_1 = 7$$

$$H_2S(aq) \rightarrow\ = 2H^+ + \rightarrow S^{2-}$$

Calculate the equilibrium constant for the overall dissociation reaction

$$HS^- = H^+ + S^{2-}; \rightarrow pK_2 = 17$$

and the concentration of each species at pH values of 3, 5, 7, 9, 11 assuming that the total dissolved $H_2S$ in solution is 0.1 M at 298K.

3. Calculate the mass of $(NH_4)_2SO_4$ you would add to a litre of water to prepare a solution of ionic strength $I = 0.5$.

4. Calculate the concentration of $HCO_3^-$ ($pK_a = 10.3$) in a solution of 0.1 M total carbonate maintained at pH 10 and 298K.

5. When a solution of 1 mol $L^{-1}$ ammonia was added drop-wise to a solution of 0.01 mol $L^{-1}$ silver nitrate at 298 K, a white precipitate was formed. As a result of the continued addition of ammonia the white precipitate redissolved forming a clear solution. How would you explain these observations? Write out relevant equations.

$$At\ 298K : AgOH(s) = Ag^+ + OH^-$$

$$Ag^+ + NH_3 = AgNH_3^+ ; Log\ K_1 = 3.31$$

$$AgNH_3^+ + NH_3 = Ag(NH_3)_2^+\ Log\ K_2 = 4.11$$

If the free ammonia concentration was maintained at 1 mol $L^{-1}$, calculate the concentration of each silver species in a clear solution containing 0.005 mol $L^{-1}$ $AgNO_3$ *in the absence of the white precipitate*.

6. For the refining of the oxide $U_3O_8$ produced in an ore treatment plant the oxide is dissolved in nitric acid. This solution is further refined by solvent extraction of $UO_2(NO_3)_2$ to remove impurities such as thorium, molybdenum, vanadium, rare earths, phosphates, arsenates, etc. The published equilibrium constants for the formation of $UO_2(NO_3)^+$ and $UO_2(NO_3)_2$ are

$$UO_2^{2+} + NO_{3-} \rightleftharpoons UO_2(NO_3)^+ : K = 0.3$$

$$UO_2^{2+} + 2NO_{3-} \rightleftharpoons UO_2(NO_3)_2^+ : K = 0.02$$

Calculate the concentration of $UO_2(NO_3)^+$ in equilibrium with $UO_2^{2+}$ in a 3 mol/L nitric acid solution. Assume that the concentration of $UO_2(NO_3)_2$ is negligibly small and the total uranium dissolved is 300 g/L (as $U_3O_8$) which exist as $UO_2^{2+}$ and $UO_2(NO_3)^+$.

7. Copper(II) forms strong complexes with ammonia, the most important of which is the tetra-ammine, $Cu(NH_3)_4^{2+}$. The stability constant ($\beta_4$) has been determined as $10^{12}$.

a. Calculate the concentration of the uncomplexed ion $Cu^{2+}$ in a solution containing 0.1 mol/L of copper ions and 2 mol/L of $NH_3$ assuming unit activity coefficients.

b. Sketch the relationship between the fraction of copper in the uncomplexed and complexed forms as a function of pH for a constant total copper and $(NH_3+NH_4)$ concentration. $pK_b$ for the protonation of $NH_3$ is 9.2.

8. Calculate the activity coefficient for $Cu^{2+}$ in a solution of 0.5 molar $CuSO_4$ (typical of the electrolyte used in the electrorefining of copper) using a) the extended Debye-Huckel equation, b) the Davies equation and c) the B-dot equation.

## Case study

The leaching (dissolution) of the important copper mineral chalcopyrite ($CuFeS_2$) in acidic solutions requires iron(III) to oxidise the mineral and thereby dissolve the copper. It has been proposed that the addition of small amounts of iodide ions can be used to increase the rate of dissolution as a result of the oxidation of iodide to iodine by the reaction

$$2Fe^{3+} + 2I^- = 2Fe^{2+} + I_2$$

with subsequent more rapid oxidation of the mineral by iodine. In iodide/iodine solutions the complexation of iodine by iodide occurs by the reaction

$$I_2(aq) + I^- = I_3^- \quad K = 835 \text{ at } 25°C.$$

This considerably complicates the interpretation of the dissolution data as a function of the iodide and iodine concentrations. To aid in the interpretation, a species diagram for the iodine/iodide/triiodide system would be useful.

## Problem

Produce a species distribution diagram for the species $I_2(aq)$, $I^-$ and $I_3^-$ in a solution of 0.01 M $I_2$ to which increasing amounts of iodide are added from 0.001 M to 0.5 M. How would the ionic strength affect this diagram.

# Appendices

## Appendix A1

**Specific heats of water at various temperatures** (Kreith, 1976).

| Temperature (°C) | $C_p^o$ (J/mol.K) |
| --- | --- |
| 25 | 75.30 |
| 50 | 75.34 |
| 75 | 75.60 |
| 100 | 75.96 |
| 150 | 77.75 |
| 200 | 81.90 |
| 250 | 87.88 |
| 300 | 105.28 |

## Appendix A2

**Specific heats of substances as functions of temperature** (Barin and Knacke, 1973).

Most entries are for solids, except where the chemical formula is followed by a (g) for gas or (l) for liquid.

The table below gives constants A, B and C for the formula:

$$C_p^o = A + B.(10^{-3}).T + C.(10^5)T^{-2}$$

where T is Kelvin.

The temperature ranges of applicability vary, but the data can be used with reasonable accuracy over the range 298K–500K and in many cases beyond this.

| Substance | A, J/(mol.K) | B | C |
| --- | --- | --- | --- |
| Ag | 23.8 | 5.1 | |
| AgCl | 63.3 | 4.2 | −11.3 |
| Ag$_2$O | 59.4 | 40.8 | −4,2 |
| Ag$_2$S | 64.6 | 40 | |
| Ag$_2$SO$_4$ | 96.7 | 116.8 | |
| Al | 20.7 | 12.4 | |
| Al$_2$O$_3$ | 103.9 | 26.3 | −29.1 |
| Al(OH)$_3$ | 30.6 | 209.9 | |
| Au | 24 | 4.4 | |
| CaO | 49.7 | 4.5 | −7.0 |

*—continued*

| Substance | A, J/(mol.K) | B | C |
|---|---|---|---|
| $Ca(OH)_2$ | 105.4 | 12 | −19.0 |
| $CaSO_4$ | 70.3 | 98.8 | |
| $CaSO_4. 0.5H_2O$ | 69.4 | 163.3 | |
| $CaSO_4.2H_2O$ | 91.4 | 318.2 | |
| Cd | 22.3 | 12.2 | |
| CdO | 48.1 | 6.4 | −4.9 |
| CdS | 54.0 | 3.8 | |
| $Cl_2(g)$ | 36.9 | 0.3 | −2.8 |
| Co | 21.5 | 13.9 | −0.8 |
| CoO | 48.3 | 8.5 | 1.7 |
| $Co_3O_4$ | 129.1 | 71.5 | −23.9 |
| $Co(OH)_2$ | 82.9 | 47.7 | |
| $CoS_{0.89}$ | 40.3 | 15.5 | |
| Cr | 19.8 | 12.9 | −0.3 |
| $Cr_2O_3$ | 119.5 | 9.2 | |
| $CrO_3$ | 82.6 | 21.7 | −17.5 |
| Cu | 24.9 | 3.8 | −1.4 |
| $Cu_2O$ | 56.6 | 29.3 | |
| CuO | 43.9 | 16.8 | −5.9 |
| $Cu(OH)_2$ | 87.0 | 23.3 | −5.4 |
| $Cu_2S$ | 81.6 | 11.1 | |
| CuS | 44.4 | — | |
| Fe | 17.5 | 24.8 | |
| FeO | 50,8 | 8.6 | −3.3 |
| $Fe_3O_4$ | 86.3 | 209.1 | |
| $Fe_2O_3$ | 98.4 | 77.9 | −14.9 |
| $Fe_2O_3 \cdot H_2O$ | 175.9 | — | |
| FeS | 0.5 | 167.5 | |
| $FeS_2$ | 74.8 | 5.6 | −12.7 |
| $H_2(g)$ | 27.3 | 3.3 | 0.5 |
| $H_2O(g)$ | 30.0 | 10.7 | 0.3 |
| $H_2O(l)$ | 69.04 | 19.7 | |
| $H_2S(g)$ | 29.4 | 15.4 | — |
| $H_2SO_4(l)$ | 157.0 | 28.3 | −23.5 |
| $H_2WO_4$ | 62.8 | 167.5 | |
| KOH | 42.7 | 77.0 | |
| Mg | 22.3 | 10.3 | −0.4 |
| Mn | 23.9 | 14.2 | −1.6 |
| MnO | 46.5 | 8.1 | −3.7 |

| | | | |
|---|---|---|---|
| Mn₃O₄ | 145.0 | 45.3 | −9.2 |
| Mn₂O₃ | 103.5 | 35.1 | −13.5 |
| MnS | 47.7 | 7.5 | — |
| N₂(g) | 27.6 | 4.3 | — |
| NH₃(g) | 29.8 | 25.1 | 1.5 |
| NaOH | 71.8 | −111.0 | — |
| Ni | 32.7 | −1.8 | −5.6 |
| NiO | -20.9 | 157.3 | 16.3 |
| NiS | 38.7 | 53.6 | — |
| NiSO₄ | 126.0 | 41.5 | — |
| O₂(g) | 30.0 | 4.2 | −1.7 |
| Pb | 23.6 | 9.7 | — |
| PbCl₂ | 60.8 | 41.6 | — |
| PbO | 41.5 | 15.3 | — |
| PbS | 46.5 | 10.3 | — |
| Pt | 24.3 | 5.4 | — |
| S | 15.0 | 26.1 | — |
| SO₂(g) | 43.5 | 10.6 | −5.9 |
| Sb | 22.4 | 9.0 | — |
| Si | 22.8 | 3.9 | −3.5 |
| SiO₂ | 43.9 | 38.8 | −9.7 |
| Sn | 21.6 | 18.2 | — |
| Ti | 22.2 | 10.3 | — |
| TiO₂ | 62.9 | 11.4 | −10.0 |
| UO₂ | 80.4 | 6.8 | −16.6 |
| U₃O₈ | 282.6 | 37.0 | −50.0 |
| UO₃ | 92.5 | 10.6 | −12.4 |
| Zn | 22.4 | 10.0 | — |
| ZnO | 49.0 | 5.1 | −9.1 |
| ZnS | 50.9 | 5.2 | −5.7 |

## Appendix A3

**Thermodynamic data for substances at 25°C** (Barin and Knacke, 1973).

| Substance | $C_p^0$(J/mol.K) | $\Delta H_f^0$(kJ/mol) | $S^0$ (J/mol.K) | $\Delta G_f^0$(kJ/mol) $\Delta G_f^0$(kJ/mol) |
|---|---|---|---|---|
| Ag | 25.4 | 0 | 42.7 | 0 |
| AgCl | 50.8 | −126.3 | 96.3 | 109.75 |

*—continued*

| Substance | $C_p^0$(J/mol.K) | $\Delta H_f^0$(kJ/mol) | $S^0$ (J/mol.K) | $\Delta G_f^0$(kJ/mol) |
|---|---|---|---|---|
| $Ag_2O$ | 66.8 | −30.6 | 121.8 | −10.82 |
| $Ag_2S$ | 76.6 | −31.8 | 143.6 | −40.26 |
| $Ag_2SO_4$ | 131.5 | −716.4 | 200.6 | −615.9 |
| Al | 24.4 | 0 | 28.3 | 0 |
| $Al_2O_3$ | 79 | −1676.5 | 51 | −1576.78 |
| $Al(OH)_3$ | 93.2 | −1285.4 | 71.2 | −1137.9 |
| Au | 25.3 | 0 | 47.5 | 0 |
| CaO | 43.2 | −634.7 | 39.8 | −604.31 |
| $Ca(OH)_2$ | 87.6 | −986.9 | 83.4 | −896.51 |
| $CaSO_4$ | 99.7 | −1435.1 | 105.3 | −1320.62 |
| $CaSO_4$ . $0.5\ H_2O$ | 118.1 | −1577.9 | 130.6 | −1435.54 |
| $CaSO_4.H_2O$ | 186.3 | −2024.1 | 194.3 | −1796.16 |
| Cd | 25.9 | 0 | 51.8 | 0 |
| CdO | 44.7 | −259.6 | 54.8 | −225.11 |
| CdS | 55.1 | −149.5 | 69.1 | −140.62 |
| $Cl_2(g)$ | 33.8 | 0 | 223.2 | 0 |
| Co | 24.8 | 0 | 30.1 | 0 |
| CoO | 52.7 | −239.1 | 53 | −205.07 |
| $Co_3O_4$ | 123.5 | −905.6 | 102.6 | −702.24 |
| $Co(OH)_2$ | 79.1 | −541.8 | 93.4 | −456.17 |
| $CoS_{0.89}$ | 44.9 | −94.6 | 52.3 | −82.86 |
| Cr | 23.3 | 0 | 23.8 | 0 |
| $Cr_2O_3$ | 104.6 | −1130.5 | 81.2 | −1047.09 |
| $CrO_3$ | 69.4 | −578.6 | 72 | (?) |
| Cu | 24.4 | 0 | 33.1 | 0 |
| $Cu_2O$ | 65.3 | −170.4 | 93 | −146.39 |
| CuO | 42.2 | −156 | 42.6 | −127.22 |
| $Cu_2S$ | 81.6 | −79.6 | 121 | −86.21 |
| CuS | 47.7 | −48.6 | 66.6 | −48.97 |
| Fe | 24.9 | 0 | 27.2 | 0 |
| FeO | 49.7 | −272.2 | 254.5 | −244.40 |
| $Fe_3O_4$ | 148.7 | −1119.2 | 146.5 | −1014.44 |
| $Fe_2O_3$ | 104.9 | −826.1 | 87.5 | 741.16 |
| $Fe_2O_3$ . $H_2O$ | 175.9 | −1118.8 | 118.9 | (?) |
| FeS | 54.7 | −95.5 | 67.4 | −97.59 |
| $FeS_2$ | 62.1 | −171.7 | 53 | −150.66 |
| $H_2(g)$ | 28.8 | 0 | 130.7 | 0 |
| $H_2O(l)$ | 75.3 | −286.1 | 70.0 | −236.9 |
| $H_2O(g)$ | 33.6 | −242.6 | 188.9 | −228.67 |

| | | | | |
|---|---|---|---|---|
| $H_2S(g)$ | 34 | −20.5 | 205.8 | −33.02 |
| $H_2SO_4(l)$ | 139 | −814.6 | 157 | −742.17 |
| KOH | 65.7 | −425 | 79.3 | −374.56 |
| Mg | 24.9 | 0 | 32.7 | 0 |
| $Mg(OH)_2$ | — | −923.8 | 63.1 | |
| Mn | 26.3 | 0 | 32 | 0 |
| MnO | 44.8 | −385.2 | 59.5 | −363.26 |
| $Mn_3O_4$ | 148.2 | −1387.6 | 154.1 | −1281.45 |
| $Mn_2O_3$ | 98.8 | −957.6 | 110.5 | −888.48 |
| MnS | 50 | −214.4 | 78.3 | −208.83 |
| $N_2(g)$ | 29.1 | 0 | 191.6 | 0 |
| $NH_3(g)$ | 53.5 | −46.1 | 192.5 | −16.64 |
| NaOH | 59.8 | −428.3 | 64.5 | −377.07 |
| Ni | 25.8 | 0 | 29.9 | 0 |
| NiO | 44.3 | −240.8 | 38.1 | −214.69 |
| NiS | 54.7 | −93 | 53 | −74.08 |
| $O_2(g)$ | 29.3 | 0 | 205.2 | 0 |
| Pb | 26.5 | 0 | 64.9 | 0 |
| $PbCl_2$ | 73.2 | −360.9 | 136.1 | −314.04 |
| PbO | 46.1 | −219.4 | 65.3 | −189.37 |
| PbS | 49.5 | −100.5 | 91.3 | −92.70 |
| Pt | 25.9 | 0 | 41.7 | 0 |
| S | 22.8 | 0 | 31.9 | 0 |
| $SO_2(g)$ | 39.9 | −297 | 248.3 | −300.44 |
| Sb | 25 | 0 | 45.6 | 0 |
| Si | 20 | 0.0 | 18.8 | 0 |
| $SiO_2$ | 44.6 | −911.5 | 41.5 | −805.19 |
| Sn | 27 | 0 | 51.6 | 0 |
| Ti | 25.2 | 0 | 30.7 | 0 |
| $TiO_2$ | 55.1 | −945.4 | 50.4 | −888.48 |
| $UO_2$ | 63.8 | −1085.7 | 77.9 | −1032.02 |
| $UO_3$ | 81.7 | −1231 | 98.8 | −1142.51 |
| Zn | 25.4 | 0 | 41.7 | 0 |
| ZnO | 40.3 | −348.4 | 43.5 | −321.3 |
| ZnS | 46.1 | −201.8 | 57.8 | −198.37 |

## Appendix A4

**Thermodynamic properties of some aqueous species at 25°C**
(Bamer and Scheuennan, 1978; Naumov et al., 1974; Zemaitis et al., 1986).

| Species | H°(kJ/mol) | $\Delta G_f^o$ (kJ/mol) | S°(J/mol.K) | $C_p^o$ (J/mol.K) |
|---|---|---|---|---|
| $Ag^+$ | 105.7 | 77.2 | 72.9 | 24.7 |
| $Al^{3+}$ | −513.7 | −485.7 | −322.0 | 38.1 |
| $Al(OH)^{2+}$ | −767.6 | −701.2 | −157.0 | — |
| $Al(OH)_2^+$ | — | −908.0 | — | — |
| $AlO_2^-$ | — | −875.6 | — | — |
| $AlF^{2+}$ | −861.4 | −803.9 | −165.4 | — |
| $AlF_2^+$ | −1191.6 | −1124.7 | −58.6 | |
| $Al(SO_4)^+$ | −1432.1 | −1251.9 | −192.2 | |
| $Al(SO_4)_2^-$ | −2338.7 | −2009.1 | −127.8 | |
| $Au^+$ | — | 163.3 | — | |
| $Au^{3+}$ | — | 433.8 | — | |
| $H_3AsO_3$ | −749.1 | −646.5 | 195.3 | 204.7 |
| $H_2AsO_3^-$ | −721.5 | −593.8 | 111.0 | −50.2 |
| $HAsO_3^-$ | −689.6 | −524.6. | −15.1 | −278.0 |
| $H_2AsO_4$ | −907.3 | −772.1 | 183.0 | 211.9 |
| $H_2AsO_4^-$ | −911.3 | −718.2 | −5.4 | −284.3 |
| $AsO_4^{-3}$ | −893.0 | −652.5 | −165.0 | −485.3 |
| $AuCl_4^-$ | −322.4 | −235.3 | 288.1 | |
| $Au(CN)_2^-$ | 242.8 | 286.0 | 192.6 | |
| $CO_3^{-2}$ | −677.6 | −528.3 | −56.9 | −250.0 |
| $HCO_3^-$ | −692.5 | −587.3 | 91.3 | −37.3 |
| $Ca^{+2}$ | −543.1 | −553.1 | −55.3 | 0.8 |
| $Cd^{+2}$ | −75.6 | −77.9 | −71.2 | 4.2 |
| $Cl^-$ | −167.3 | −131.4 | 56.5 | −136.1 |
| $Co^{+2}$ | −58.6 | −56.1 | −108.9 | 37.3 |
| $Cr^{+2}$ | −139.0 | −176.3 | — | |
| $Cr^{+3}$ | −221.9 | −204.1 | −316.5 | 49.4 |
| $Cu^+$ | 72.2 | 50.0 | 42.3 | 59.9 |
| $Cu^{+2}$ | 65.7 | 65.3 | 95.9 | 23.4 |
| $CuCl_2^-$ | — | −240.3 | — | |
| $CuSO_4$ | −200.4 | −691.9 | −18.0 | |
| $Fe^{+2}$ | −89.2 | −78.9 | −137.8 | 33.1 |
| $Fe^{+3}$ | −48.6 | −4.6 | −316.1 | 24.7 |
| $FeOH^+$ | −324.9 | −277.6 | −29.3 | — |
| $Fe(OH)_2^{+2}$ | −291.0 | −229.6 | −142.4 | — |
| $FeCl^{+2}$ | −180.5 | −144.0 | −113.0 | — |

| | | | |
|---|---|---|---|
| $FeSO_4$ | −999.0 | −832.5 | −117.7 | — |
| $FeSO_4^+$ | −932.4 | −773.3 | −129.8 | — |
| $Mg^{+2}$ | −467.2 | −445.1 | −138.2 | 129.4 |
| $Mn^{+2}$ | −220.9 | −228.2 | −73.7 | 50.2 |
| $MnSO_4$ | −1128.0 | −985.3 | 2.5 | — |
| $MnO4^-$ | −533.4 | −440.6 | 196.4 | −108.4 |
| $NH_4^+$ | −132.6 | −79.4 | 114.7 | 80.8 |
| $NH_3$ | −80.3 | −26.6 | 111.4 | — |
| $NO_3^-$ | −207.5 | −111.4 | 146.5 | 86.7 |
| $Na^+$ | −240.3 | −262.1 | 59.0 | 46.5 |
| $Ni^{+2}$ | −54.0 | −45.6 | −129.0 | 5.9 |
| $Ni(NH_3)_2^{+2}$ | −246.6 | −128.1 | 85.4 | — |
| $Ni(NH_3)_4^{+2}$ | −439.2 | — | 258.8 | — |
| $Ni(NH_3)_6^{+2}$ | −630.6 | −256.2 | 394.8 | — |
| $NiSO_4$ | −961.8 | −802.8 | 61.1 | — |
| $OH^-$ | −230.2 | −157.4 | −10.8 | −148.6 |
| $Pb+2$ | −1.7 | −24.4 | 11.3 | −53.6 |
| $PbCl^+$ | — | −165.0 | — | — |
| $PtCl_4^{-2}$ | −503.3 | −368.9 | 167.5 | — |
| $PtCl_6^{-2}$ | −675.4 | −490.0 | 220.6 | — |
| $S^{-2}$ | 33.1 | 85.8 | −14.7 | −400.7 |
| $HS^-$ | −17.6 | 12.1 | 62.8 | −140.7 |
| $H_2S$ | −39.8 | 27.9 | 121.4 | 131.9 |
| $SO_4^{2-}$ | −909.9 | −744.4 | 20.1 | −293.1 |
| $HSO_4^-$ | −888.0 | −756.5 | 131.9 | −83.7 |
| $UO_2^{2+}$ | −1025.0 | −962.1 | −86.3 | 39.4 |
| $Zn^{2+}$ | −154.0 | −147.1 | −112.2 | 46.1 |

# Appendix A5

**Debye−Huckel constants for water at 1 atmosphere** (Bamer and Scheuennan, 1978).

| Temperature (°C) | A(kg$^{1/2}$ mol$^{-1/2}$) | B (kg$^{1/2}$ cm mol$^{-1/2}$)× 10$^{-8}$ |
|---|---|---|
| 0 | 0.4883 | 0.3241 |
| 5 | 0.4921 | 0.3249 |
| 10 | 0.4960 | 0.3258 |
| 15 | 0.5000 | 0.3262 |
| 20 | 0.5042 | 0.3273 |
| 25 | 0.5085 | 0.3281 |

*—continued*

| Temperature (°C) | $A(kg^{1/2}\ mol^{-1/2})$ | $B\ (kg^{1/2}\ cm\ mol^{-1/2})\times 10^{-8}$ |
|---|---|---|
| 30 | 0.5130 | 0.3290 |
| 35 | 0.5175 | 0.3297 |
| 40 | 0.5221 | 0.3305 |
| 45 | 0.5271 | 0.3314 |
| 50 | 0.5319 | 0.3321 |
| 55 | 0.5371 | 0.3329 |
| 60 | 0.5425 | 0.3338 |

## Appendix A6

**Some values of the parameter $a_i$** (Huheey et al., 2006).

| $a_i \times 10^8$ (cm) | Ions |
|---|---|
| 2.5 | $NH_4^+$, $Ag^+$, $Tl^+$ |
| 3.0 | $K^+$, $Cl^-$, $Br^-$, $I^-$, $CN^-$, $NO_3^-$ |
| 3.5 | $OH^-$, $F^-$, $HS^-$, $CNS^-$, $MnO_4^-$ |
| 4.0 | $Na^+$, $Hg_2^{2+}$, $HCO_3^-$, $SO_4^{2-}$, $S_2O_3^{2-}$, $PO_4^{3-}$ |
| 4.5 | $Pb^{2+}$, $CO_3^{2-}$, $SO_3^{2-}$, $MoO_4^{2-}$ |
| 5.0 | $Sr^{2+}$, $Ba^{2+}$, $Cd^{2+}$, $Hg^{2+}$, $S^{2-}$, $S_2O_4^{2-}$, $WO_4^{2-}$, |
| 6 | $Li^+$, $Ca^{2+}$, $Cu^{2+}$, $Zn^{2+}$, $Mn^{2+}$, $Fe^{2+}$, $Ni^{2+}$, $Co^{2+}$, $Sn^{2+}$ |
| 8 | $Mg^{2+}$, $Be^{2+}$ |
| 9 | $H^+$, $Al^{3+}$, $Cr^{3+}$, $Fe^{3+}$, $Ce^{3+}$, La(other rare earth ions)$^{3+}$ |
| 11 | $Th^{4+}$, $Zr^{4+}$, $Ce^{4+}$, $Sn^{4+}$ |

## Appendix A7

**Criss and Cobble Parameters**(Criss and Cobble, 1964).

The values of $\alpha(T)$ which has units of J/mol.K and $\beta(T)$ which is dimensionless are given for various types of ions.

Simple anions are $OH^-$, $Cl^-$, oxy - anions include $SO_4^{2-}$, $CO_3^{2-}$, $NO_3^-$ and acid oxy - anions are of the type $HSO_4^-$, $HCO_3^-$, $H_2PO_4^-$.

As the absolute entropy of the $H^+$ ion is selected in order to obtain a linear relationship in accordance with Eq. (2.27), its average $C_P^o$ is presented in a separate column.

| T(K) | H⁺ | Cations | | Simple | Anions |
|------|------|------|------|------|------|
| | $C_p^o(T)$ | $\alpha(T)$ | $\beta(T)$ | $\alpha(T)$ | $\beta(T)$ |
| 333 | 96 | 147 | −0.41 | −192 | −0.28 |
| 373 | 130 | 192 | −0.55 | −242 | 0 |
| 423 | 138 | 194 | −0.59 | −255 | −0.03 |
| 473 | 147 | 211 | −0.63 | −274 | −0.04 |
| | H⁺ | Oxy− | anions | Acid oxy− | anions |
| | $S^o(T)$ | $\alpha(T)$ | $\beta(T)$ | $\alpha(T)$ | $\beta(T)$ |
| 298 | −20.9 | | | | |
| 333 | −10.5 | −528 | 1.96 | −509 | 3.42 |
| 373 | 8.4 | −578 | 2.12 | −565 | 3.98 |
| 423 | 27.2 | −554 | 1.96 | −597 | 3.94 |
| 473 | 46.4 | −607 | 2.21 | −634 | 4.24 |

## Appendix A8

**Solubility products for sparingly soluble compounds** (Martell and Smith, 1976).

| Reaction | $Log_{10}K_{sp}$ (25°C) |
|------|------|
| $AgCl(s) = Ag^+ + Cl^-$ | −9.74 |
| $Ag_2SO_4(s) = 2Ag^+ + SO_4^{-2}$ | −4.83 |
| $Al(OH)_3(s) = Al^{+3} + 3OH^-$ | −32.34 |
| $CaCO_3(s) = Ca^{+2} + CO_3^{-2}$ | −8.35 |
| $Ca(OH)_2(s) = Ca^{+2} + 2OH^-$ | −5.19 |
| $CaSiO_3(s) = Ca^{+2} + SiO_3^{-2}$ | −7.2 |
| $CaSO_4(s) = Ca^{+2} + SO_4^{-2}$ | −4.62 |
| $Cd(OH)_2(s) = Cd^{+2} + 2OH^-$ | −14.35 |
| $CdS(s) = Cd^{+2} + S^{-2}$ | −27 |
| $Co(OH)_2(s) = Co^{+2} + 2OH^-$ | −14.9 |
| $CoS(s) = Co^{+2} + S^{-2}$ | −21.3 |
| $Cu_2S(s) = 2Cu^+ + S^{-2}$ | −48.5 |
| $Cu(OH)_2(s) = Cu^{+2} + 2OH^-$ | −19.32 |
| $CuS(s) = Cu^{+2} + S^{-2}$ | −36.1 |
| $Fe(OH)_2(s) = Fe^{+2} + 2OH^-$ | −15.1 |
| $FeS(s) = Fe^{+2} + S^{-2}$ | −18.1 |
| $Fe(OH)_3(s) = Fe^{+3} + 3OH^-$ | −38.8 |
| $FeOOH(s) + H_2O = Fe^{+3} + 3OH^-$ | −39.5 |
| $Mg(OH)_2(s) = Mg^{+2} + 2OH^-$ | −11.15 |

*—continued*

| Reaction | $Log_{10}K_{sp}$ (25°C) |
|---|---|
| $MnCO_3(s) = Mn^{+2} + CO_3^{-2}$ | −9.3 |
| $Mn(OH)_2(s) = Mn^{+2} + 2OH^-$ | −12.8 |
| $MnS(s) = Mn^{+2} + S^{-2}$ | −10.5 |
| $NiCO_3(s) = Ni^{+2} + CO_3^{-2}$ | −6.87 |
| $Ni(OH)_2(s) = Ni_{+2} + 2OH^-$ | −15.2 |
| $NiS(s) = Ni^{+2} + S^{-2}$ | −19.4 |
| $PbCl_2(s) = Pb^{+2} + 2Cl^-$ | −4.78 |
| $PbCO_3(s) = Pb^{+2} + CO_3^{-2}$ | −13.13 |
| $PbS(s) = Pb^{+2} + S^{-2}$ | −27.5 |
| $PbSO_4(s) = Pb^{+2} + SO_4^{-2}$ | −7.79 |
| $UO_2(OH)_2(s) = UO_2^{+2} + 2OH^-$ | −22.4 |
| $ZnCO_3(s) = Zn^{+2} + CO_3^{-2}$ | −10.0 |
| $am\text{-}Zn(OH)_2(s) = Zn^{+2} + 2OH^-$ | −15.52 |
| $ZnS(s) = Zn^{+2} + S^{-2}$ | −22.5 |

# Appendix A9

**Equilibrium constants at 25°C for reactions involving aqueous species** (Martell and Smith, 1976).

| Reaction | $Log_{10}K$ |
|---|---|
| *Chloride Complexes* | |
| $Au^+ + 2Cl^- = AuCl_2^-$ | 9.7 |
| $Au^{3+} + 4Cl^- = AuCl_4^-$ | 12(?) |
| $Ag^+ + Cl^- = AgCl$ | 3.31 |
| $Ag^+ + 2Cl^- = AgCl_2^-$ | 5.25 |
| $Cd^{+2} + Cl^- = CdCl^+$ | 1.98 |
| $Cd^{+2} + 2Cl^- = CdCl_2$ | 2.6 |
| $Cd^{+2} + 3Cl^- = CdCl_3^-$ | 2.4 |
| $Cd^{+2} + 4Cl^- = CdCl_4^{-2}$ | 1.7 |
| $Cu^+ + 2Cl^- = CuCl_2^-$ | 5.8 |
| $Cu^{+2} + Cl^- = CuCl^+$ | 0.4 |
| $Cu^{+2} + 2Cl^- = CuCl^2$ | −0.4 |
| $Fe^{+3} + Cl^- = FeCl^{+2}$ | 1.48 |
| $Fe^{+3} + 2Cl^- = FeCl_2^+$ | 2.13 |
| $Fe^{+3} + 3Cl^- = FeCl_3$ | 1.1 |
| $K^+ + Cl^- = KCl$ | −0.5 |

| | |
|---|---|
| $Na^+ + Cl^- = NaCl$ | −1.85 |
| $Pb^{+2} + Cl^- = PbCl^+$ | 1.59 |
| $Pb^{+2} + 2Cl^- = PbCl_2$ | 1.8 |
| $Pb^{+2} + 3Cl^- = PbCl_3^-$ | 1.7 |
| $Pb^{+2} + 4Cl^- = PbCl_4^{-2}$ | 1.4 |
| $Zn^{+2} + Cl^- = ZnCl^+$ | 0.43 |
| $Zn^{+2} + 2Cl^- = ZnCl_2$ | 0.61 |
| $Zn^{+2} + 3Cl^- = ZnCl_3^-$ | 0.5 |
| $Zn^{+2} + 4Cl^- = ZnCl_4^{-2}$ | 0.2 |

## Cyanide Complexes

| | |
|---|---|
| $Au^+ + 2CN^- = Au(CN)_2^-$ | 38.8 |
| $Ag^+ + 2CN^- = Ag(CN)_2^-$ | 20.5 |
| $Cd^{+2} + 4CN^- = Cd(CN)_4^{-2}$ | 17.9 |
| $Cu^+ + 2CN^- = Cu(CN)_2^-$ | 23.1 |
| $Fe^{+2} + 6CN^- = Fe(CN)_6^{-4}$ | 35.4 |
| $Fe^{+3} + 6CN^- = Fe(CN)_6^{-3}$ | 43.6 |
| $H^+ + CN^- = HCN(a)$ | 9.2 |
| $Mn^{+3} + 6CN^- = Mn(CN)_6^{-3}$ | 9.7 |
| $Ni^{+2} + 4CN^- = Ni(CN)_4^{-2}$ | 30.2 |
| $Zn^{+2} + 4CN^- = Zn(CN)_4^{-2}$ | 19.6 |

## Ammonia Complexes

| | |
|---|---|
| $Ag^+ + NH_3 = AgNH_3^+$ | 3.31 |
| $Ag^+ + 2NH_3 = Ag(NH_3)_2^+$ | 7.22 |
| $Cu^{+2} + NH_3 = CuNH_3^{+2}$ | 4.04 |
| $Cu^{+2} + 2NH_3 = Cu(NH_3)_2^{+2}$ | 7.47 |
| $Cu^{+2} + 3NH_3 = Cu(NH_3)_3^{+2}$ | 10.27 |
| $Cu^{+2} + 4NH_3 = Cu(NH_3)_4^{+2}$ | 11.75 |
| $H^+ + NH_3 = NH_4^+$ | 9.24 |
| $Mn^{+2} + NH_3 = MnNH_3^{+2}$ | 1.00 |
| $Ni^{+2} + NH_3 = NiNH_3^{+2}$ | 2.72 |
| $Ni^{+2} + 2NH_3 = Ni(NH_3)_2^{+2}$ | 4.89 |
| $Ni^{+2} + 3NH_3 = Ni(NH_3)_3^{+2}$ | 6.55 |
| $Ni^{+2} + 4NH_3 = Ni(NH_3)_4^{+2}$ | 7.67 |
| $Ni^{+2} + 5NH_3 = Ni(NH_3)_5^{+2}$ | 8.34 |
| $Ni^{+2} + 6NH_3 = Ni(NH_3)_6^{+2}$ | 8.31 |
| $Zn^{+2} + NH_3 = ZnNH_3^{+2}$ | 2.21 |
| $Zn^{+2} + 2NH_3 = Zn(NH_3)_2^{+2}$ | 4.50 |
| $Zn^{+2} + 3NH_3 = Zn(NH_3)_3^{+2}$ | 6.86 |
| $Zn^{+2} + 4NH_3 = Zn(NH_3)_4^{+2}$ | 8.89 |

*—continued*

| Reaction | $Log_{10}K$ |
|---|---|
| **Hydroxide Complexes** | |
| $Al^{+3} + OH^- = AlOH^{+2}$ | 9.03 |
| $Al^{+3} + 2OH^- = Al(OH)_2^+$ | 18.69 |
| $Al^{+3} + 3OH^- = Al(OH)_3$ | 26.99 |
| $Al^{+3} + 4OH^- = Al(OH)_4^-$ | 32.99 |
| $2Al^{+3} + 2OH^- = Al_2(OH)_2^{+4}$ | 20.30 |
| $Ca^{+2} + OH^- = CaOH^+$ | 1.15 |
| $Co^{+2} + OH^- = CoOH^+$ | 4.35 |
| $Co^{+2} + 2OH^- = Co(OH)_2$ | 9.19 |
| $Cr^{+3} + OH^- = CrOH^{+2}$ | 10.0 |
| $Cr^{+3} + 2OH^- = Cr(OH)_2^+$ | 18.3 |
| $Cr^{+3} + 3OH^- = Cr(OH)_3$ | 24 |
| $Cr^{+3} + 4OH^- = Cr(OH)_4^-$ | 28.6 |
| $Cu^{+2} + OH^- = CuOH^+$ | 6.3 |
| $Cu^{+2} + 2OH^- = Cu(OH)_2$ | 10.7 |
| $Cu^{+2} + 3OH^- = Cu(OH)_3^-$ | 14.2 |
| $Cu^{+2} + 2OH^- = Cu(OH)_4^{-2}$ | 16.4 |
| $Fe^{+2} + OH^- = FeOH^+$ | 4.5 |
| $Fe^{+2} + 2OH^- = Fe(OH)_2$ | 7.4 |
| $Fe^{+2} + 3OH^- = Fe(OH)_3^-$ | 11 |
| $Fe^{+2} + 2OH^- = Fe(OH)_4^{-2}$ | 10 |
| $Fe^{+3} + OH^- = FeOH^{+2}$ | 11.81 |
| $Fe^{+3} + 2OH^- = Fe(OH)_2^+$ | 22.32 |
| $Fe^{+3} + 3OH^- = Fe(OH)_3$ | 30 |
| $Fe^{+3} + 4OH^- = Fe(OH)_4^-$ | 34.4 |
| $2Fe^{+3} + 2OH^- = Fe_3(OH)_2^{+4}$ | 25.04 |
| $H^+ + OH^- = H_2O$ | 14.00 |
| $Mg^{+2} + OH^- = MgOH^+$ | 2.55 |
| $Mn^{+2} + OH^- = MnOH^+$ | 3.40 |
| $Mn^{+2} + 2OH^- = Mn(OH)_2$ | 5.8 |
| $Pb^{+2} + OH^- = PbOH^+$ | 6.29 |
| $Pb^{+2} + 2OH^- = Pb(OH)_2$ | 10.87 |
| $UO2^{+2} + OH^- = UO_2OH^+$ | 8.19 |
| $Zn^{+2} + OH^- = ZnOH^+$ | 5.04 |
| $Zn^{+2} + 2OH^- = Zn(OH)_2$ | 11.09 |
| $Zn^{+2} + 3OH^- = Zn(OH)_3^-$ | 13.52 |
| $Zn^{+2} + 4OH^- = Zn(OH)_4^{-2}$ | 14.8 |

## Sulphate Complexes

| | |
|---|---|
| $Al^{+3} + SO_4^{-2} = AlSO_4^+$ | 3.89 |
| $Ag^+ + SO_4^{-2} = AgSO_4(-)$ | 1.3 |
| $Ca^{+2} + SO_4^{-2} = CaSO_4$ | 2.31 |
| $Ce^{+3} + SO_4^{-2} = CeSO_4^+$ | 3.59 |
| $Cd+2 + SO_4^{-2} = CdSO_4$ | 2.46 |
| $Co^{+2} + SO_4^{-2} = CoSO_4$ | 2.36 |
| $Cu^{+2} + SO_4^{-2} = CuSO_4$ | 2.36 |
| $Fe^{+2} + SO_4^{-2} = FeSO_4$ | 2.2 |
| $Fe^{+3} + SO_4^{-2} = FeSO_4^+$ | 4.04 |
| $Fe^{+3} + 2SO_4^{-2} = Fe(SO_4)_2^-$ | 5.38 |
| $H^+ + SO_4^{-2} = HSO_4^-$ | 1.91 |
| $K^+ + SO_4^{-2} = KSO_4^-$ | 0.85 |
| $Mg^{+2} + SO_4^{-2} = MgSO_4$ | 2.23 |
| $Mn^{+2} + SO_4^{-2} = MnSO_4$ | 2.26 |
| $Na^+ + SO_4^{-2} = NaSO_4^-$ | 0.70 |
| $NH_4^+ + SO_4^{-2} = NH_4SO_4^-$ | 1.95 |
| $Pb^{+2} + SO_4^{-2} = PbSO_4$ | 2.75 |
| $UO_2^{+2} + SO_4^{-2} = UO_2SO_4$ | 2.95 |
| $Zn^{+2} + SO_4^{-2} = ZnSO_4$ | 2.38 |

# 3

# Chemical equilibria in hydrometallurgical reactions

In Chapter 2, the concepts of hard and soft acids and bases, the hydrolysis of metal ions, the formation and stability of soluble complexes, and their stability constants were introduced together with some aspects of the thermodynamics of ions in solution. The focus has been on homogeneous systems. In this chapter, we will further develop the concepts required for an understanding of the thermodynamics, but in this case, we shall focus on the equilibria involved in the heterogeneous chemical reactions taking place in hydrometallurgical processes. A knowledge of the equilibrium in any process such as leaching, precipitation, separation by liquid-liquid or solid-liquid extraction, and recovery by reduction is a necessary condition for successful implementation and operation.

The following are some of the types of reactions which are encountered in hydrometallurgical processes together with some examples.

- **Acid-base reactions**

$$ZnO(s) + H_2SO_4 = ZnSO_4(a) + H_2O$$

$$Al_2O_3(s) + 2NaOH + 3H_2O = 2Al(OH)_4^- + 2Na^+$$

- **Complexation reactions**

$$Au^+ + 2CN^- = Au(CN)_2^-$$

$$Co^{2+} + 5NH_3 = Co(NH_3)_5^{2+}$$

- **Precipitation reactions**

$$Pb^{2+} + S^{2-} = PbS(s)$$

$$Ni^{2+} + 2OH^- = Ni(OH)_2(s)$$

- **Redox reactions**

$$Ni^{2+} + 2e = Ni(s)$$

Hydrometallurgy. https://doi.org/10.1016/B978-0-323-99322-7.00007-7

$$NiS(s) + 2Fe^{3+} = Ni^{2+} + 2Fe^{2+} + S$$

$$Cu^{2+} + Zn = Cu(s) + Zn^{2+}$$

In many cases, the position of equilibrium is so far to the right, that is, the free energy change is large and negative, that we do not have to take the thermodynamics into account in a quantitative description of the reaction or process. However, in a significant number of instances this is not the case, and a knowledge of the thermodynamics is essential.

## 3.1 Gibbs free energy change for a reaction

In order to determine the equilibrium conditions for a reaction such as

$$aA + bB \rightleftharpoons cC \tag{3.1}$$

it is necessary to calculate the Gibbs free energy change, which for the above reaction is:

$$\Delta G_R(T) = \Delta G_R^o(T) + RT \ln \frac{\left(a_C/a_C^o\right)^c}{\left(a_A/a_A^o\right)^a \left(a_B/a_B^o\right)^b} \tag{3.2}$$

in which $\Delta G_R^o(T)$ is the standard Gibbs free energy change for the reaction, $a_A$, $a_B$ and $a_C$ are the activities for each of the species and $a_A^o$, $a_B^o$ and $a_C^o$ are the corresponding activities in the standard state. At equilibrium, $\Delta G_R(T) = 0$ and

$$\frac{\left(a_C/a_C^o\right)^c}{\left(a_A/a_A^o\right)^a \left(a_B/a_B^o\right)^b} = \exp\left(\frac{-\Delta G_R^o(T)}{RT}\right) = K \tag{3.3}$$

in which K is the equilibrium constant for the reaction. Note that K is dimensionless because the activities are all measured relevant to the standard state.

The standard free energy change for the reaction at temperature T is the sum of the free energies of formation of the products minus the free energies of formation of the reactants, all in their standard states, that is

$$\Delta G_R^o(T) = c\Delta G_{fC^o}(T) - a\Delta G_{fA^o}(T) - b\Delta G_{fB^o}(T)$$
$$= -RT \ln K \tag{3.4}$$

Values for $\Delta G_f^o(298)$ are given in Appendix A4 of Chapter 2 for a number of important species.

Note that a reaction will occur spontaneously if $\Delta G < 0$ or, in equivalent terms, $K > 1$.

Similarly, one can calculate the enthalpy and entropy changes for the above reaction

$$\Delta H_R^o(T) = c\Delta H_{fC}^o(T) - a\Delta H_{fA}^o(T) - b\Delta H_{fB}^o(T) \qquad (3.5)$$

$$\Delta S_R^o(T) = cS_C^o(T) - aS_A^o(T) - bS_B^o(T)$$

These quantities are related by the well-known equation

$$\Delta G^o = \Delta H^o - T\Delta S^o \qquad (3.6)$$

which can be used to estimate $\Delta G$ values from tabulations of enthalpy and entropy data.

Estimate the enthalpy (per mole Cu produced) associated with the electrowinning of copper at 298K and 1 atmosphere pressure by the reaction

$$CuSO_4(a) + H_2O(l) = Cu(s) + H_2SO_4(a) + 1/2O_2(g)$$

## 3.2 Equilibria involving hydrolysis

In Section 2.4, the hydrolysis of divalent metal ions was introduced. Both mono- and polynuclear species may be formed by hydrolysis of metal ions. For example, in acidic solutions, $Fe^{3+}$ forms a binuclear species $Fe_2(OH)_2^{4+}$ and two mononuclear species $FeOH^{2+}$ and $Fe(OH)_2^{+}$, with the former becoming more prevalent at higher concentrations.

The various stages of hydrolysis for mononuclear species may in most cases be written as

$$M^{n+} + mH_2O \rightleftharpoons M(OH)_m^{(n-m)+} + mH^+ \qquad (3.7)$$

with $M^{n+}$ being the metal ion with charge $n+$ and $m$ being an integer of one and higher. All the components in the reaction will be taken to be in the aqueous phase, unless otherwise specified. Thus $M(OH)_m^{(n-m)+}$ will be a positive ion when $n > m$, neutral soluble species when $n = m$ and a negative ion when $n < m$. The equilibrium constant for the above reaction is given by

$$K_m = \frac{(M(OH)_m^{(n-m)+})(H^+)^m}{(M^{n+})(H_2O)^m} \qquad (3.8)$$

Because of the large range of values for $K_m$ for various metal ions, the data are often reported as logarithms to the base ten, that is:

$$\log K_m = \log\left(M(OH)_m^{(n-m)+}\right) - \log(M^{n+}) - m.\log(H_2O) - m.pH$$

(3.9)

In many instances when $m = n$, the solubility of the neutral species is very low. In such cases, the product of hydrolysis is likely to be a solid precipitate, indicated by (s) after the species,

$$M^{n+} + nH_2O \rightleftharpoons M(OH)_n(s) + nH^+$$

(3.10)

For such cases, the equilibrium constant will be denoted by $K_s$, that is:

$$K_s = \frac{[H^+]^n}{[M^{n+}][H_2O]^n}$$

(3.11)

Note that the activity of the solid product is omitted as its value is likely to be close to unity if it is the pure material and the activity of water can be taken as unity.

Data for sparingly soluble species are often recorded as **solubility products**, which for the reaction

$$M^{n+} + nOH^- \rightleftharpoons M(OH)_n(s)$$

(3.12)

are defined by

$$K_{sp} = (M^{n+})(OH^-)^n$$

(3.13)

Solubility product data are given in the Appendix for a number of sparingly soluble salts.

Of particular interest in hydrometallurgy are divalent cations. Taking magnesium as an example, the first stage of hydrolysis is:

$$Mg^{2+} + H_2O \rightleftharpoons MgOH^+ + H^+$$

(3.14)

$$\log K_1 = -11.44 (\text{at } 25°C)$$

(3.15)

At a pH of 8.4 with $(Mg^{2+}) = 1$ and $(H_2O) = 1$,

$$(MgOH^+) = 10^{-3}$$

If the pH increases any higher, $Mg(OH)_2(s)$ will precipitate, thus it can be concluded that $MgOH^+$ does not contribute significantly to the overall concentration of magnesium in solution. The second hydrolysis constant has not been reported in the literature as it is apparently immeasurably low and beyond the capabilities of existing experimental methods. The $Mg^{2+}$ ion is therefore the dominant ion over the whole range of useful pH values.

Solid-liquid equilibrium between hydroxides of the form $M(OH)_2(s)$ are important in at least two areas of hydrometallurgy. First, selective precipitation may permit separation of certain metal hydroxides by the correct choice of pH. Secondly, in leaching metal oxides from an ore, the solid-liquid equilibrium is likely to be controlled by the formation of a surface hydroxide when the ore is in contact with water, i.e.:

$$MgO(s) + H_2O = Mg(OH)_2(s) \qquad (3.16)$$

The pH at which leaching takes place will determine the likely equilibrium concentration of the $Mg^{2+}$ ion in solution. In a mixed oxide ore, it may be possible to selectively leach certain components by control of the pH of the solution.

The above effects can best be illustrated graphically by rearranging the equation for the solubility product for n = 2 and setting $[H_2O] = 1$, that is,

$$\log(M^{2+}) = \log K_{sp} + 2.\log K_w - 2.pH \qquad (3.17)$$

This is plotted in Fig. 3.1 for several metal ions. Note that the line for cobalt (not shown) lies very close to that of nickel.

A calcined ore contains MnO and FeO. On leaching the ore, an equilibrium $[Mn^{2+}]$ of 0.4 mol dm$^{-3}$ is obtained at 25°C. What is the pH in the leach tank and the equilibrium $[Fe^{2+}]$?

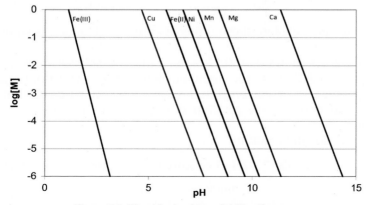

**Figure 3.1** Metal hydroxide solubility diagram.

Similar diagrams can be drawn for other compounds, the most important group of which is the sulphides in which case the additional line for the equilibrium

$$H_2S(aq) = H^+ + HS^-  \tag{3.18}$$

can also be drawn on the same diagram and this enables one to make predictions of the solubility of metal ions as a function of both sulphide activity and pH. We will return to the application of these and other solubility diagrams in later Chapters.

Hydrolysis beyond the second stage sometimes results in a significant redissolution of the hydroxide precipitate. This occurs in the case of zinc due to the formation of hydroxy—anions with a higher solubility than those of other divalent metals, namely, $Zn(OH)_3^-$ and $Zn(OH)_4^{2-}$. The equilibrium hydrolysis constants (Eq. 3.8) for $Zn^{2+}$ are:

$$\log K_1 = -8.96, \log K_s = -16.9, \log K_3 = -28.4 \text{ and } \log K_4 = -41.2$$

Defining $R_1$, $R_2$, $R_3$ and $R_4$, that is the log of the concentration fraction of the species relative to the aquo-ion as:

$$R_1 = \log \frac{(ZnOH^+)}{(Zn^{2+})} = -8.96 + pH  \tag{3.19}$$

$$R_2 = \log \frac{(Zn(OH)_2)}{(Zn^{2+})} = -16.9 + 2pH  \tag{3.20}$$

$$R_3 = \log \frac{(Zn(OH)_3^-)}{(Zn^{2+})} = -28.4 + 3pH  \tag{3.21}$$

$$R_4 = \log \frac{(Zn(OH)_4^{2-})}{(Zn^{2+})} = -41.2 + 4pH  \tag{3.22}$$

One can calculate these ratios at various pH values as shown in Table 3.1.

Table 3.1 shows that $Zn^{2+}$ is the dominant species at low pH, whereas $Zn(OH)_4^{2-}$ is the dominant ion at high pH. In the mid-pH range $Zn(OH)_2(s)$ precipitates with the result that the total zinc concentration in solution is likely to be very low in this region.

Thus, for the reaction

$$Zn^{2+} + 2H_2O \rightleftharpoons Zn(OH)_2(s) + 2H^+  \tag{3.23}$$

$$\log K_s = -11.0  \tag{3.24}$$

**Table 3.1 Calculated fractions(logs) of zinc species concentrations relative to the aquo-ion.**

| pH | $R_1$ | $R_2$ | $R_3$ | $R_4$ |
|----|-------|-------|-------|-------|
| 4  | −5.0  | −8.9  | −16.4 | −25.2 |
| 6  | −3.0  | −4.9  | −10.4 | −17.2 |
| 8  | −1.0  | −0.9  | −4.4  | −9.2  |
| 10 | 1.0   | 3.1   | 1.6   | −1.2  |
| 12 | 3.0   | 7.1   | 4.6   | 6.8   |
| 14 | 5.0   | 11.1  | 13.6  | 14.8  |

$$\log(Zn^{2+}) = -\log K_s - 2pH \qquad (3.25)$$

Thus for $(Zn^{2+}) = 1$, the equilibrium pH is 5.5.
Similarly,

$$Zn(OH)_4^{2-} + 2H^+ \rightleftharpoons Zn(OH)_2(s) + 4H_2O \qquad (3.26)$$

$$\text{Log } K = 29.8 \qquad (3.27)$$

and, for $(Zn(OH)_4^{2-}) = 1$, the equilibrium pH is 14.9.

Hydrolysis products of the trivalent ion, $Fe^{3+}$ are of particular interest in hydrometallurgy, as iron oxides are present in practically all types of mined ore. One of the most common ways of controlling the iron content of a solution is by precipitation of ferric hydroxides in the form of $Fe(OH)_3(s)$, $FeOOH(s)$ (goethite) or $Fe_2O_3(s)$ (hematite).

The soluble hydrolysis products of $Fe^{3+}$ have been widely studied and the equilibrium constants at 25°C of the more relevant ones are:

$$Fe^{3+} + H_2O \rightleftharpoons FeOH^{2+} + H^+ \quad LogK_1 = -2.19 \qquad (3.28)$$

$$Fe^{3+} + 2H_2O \rightleftharpoons Fe(OH)_2^+ + 2H^+ \quad LogK_2 = -5.67 \qquad (3.29)$$

$$2Fe^{3+} + 2H_2O \rightleftharpoons Fe_2(OH)_2^{4+} + 2H^+ \quad LogK_{22} = -2.95 \qquad (3.30)$$

The solid-liquid equilibrium reaction and the equilibrium constant at 25°C are given by:

$$Fe^{3+} + 3H_2O \rightleftharpoons Fe(OH)_3(s) + 3H^+ \quad logK_s = -3.2 \qquad (3.31)$$

From these, the total Fe(III) solution activity in equilibrium with the various solid products can be determined.

## Determine the solubility (in terms of the total Fe(III) activity) of Fe(OH)$_3$(s) as a function of pH.

Solution:

Let us work out the aquo-ion concentration first from the solubility product. For the reaction:

$$Fe^{3+} + 3H_2O \rightleftharpoons Fe(OH)_3(s) + 3H^+$$

$K_{sp} = \frac{[H^+]^3}{[Fe^{3+}]}$ and, using concentrations instead of activities

$$\log[Fe^{3+}] = 3.2 - 3pH$$

$$Fe^{3+} + H_2O \rightleftharpoons FeOH^{2+} + H^+$$

$$\frac{\left[FeOH^{2+}\right]\left[H^+\right]}{Fe^{3+}} = 10^{-2.19}$$

$$\log\left[FeOH^{2+}\right] = \log\left[Fe^{3+}\right] + pH - 2.19$$

$$= 1.01 - 2pH$$

Similarly,

$$\log[Fe(OH)_2^+] = -2.47 - pH$$

$$\log[Fe_2(OH)_2^{4+}] = 3.45 - 4pH$$

Total $[Fe(III)] = [Fe^{3+}] + [FeOH^{2+}] + [Fe(OH)_2^+] + 2[Fe_2(OH)_2^{4+}]$ and substituting in the above equations we can derive the concentrations plotted in Fig. 3.2.

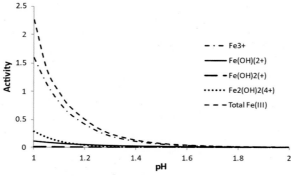

**Figure 3.2** Effect of pH on concentration of different iron(III) species in equilibrium with Fe(OH)$_3$(s).

It is clear that the total iron concentration in solution decreases rapidly as the pH increases and becomes close to zero at pH values above about 2. The aqua-ion $Fe^{3+}$ is the dominant soluble iron(III) species at low pH values, while the concentration of the dimeric species is greater than that of the other hydroxy-complex ions. We shall return in more detail to these precipitation reactions in Chapter 10.

## 3.3 Effect of temperature on aqueous equilibria

This topic was introduced in Section 2.12 as applied to homogeneous equilibria and reactions. The development of industrial processes such as high temperature pressure leaching of sulphide and oxide ores and mattes requires accurate thermodynamic data for ions at elevated temperatures. However, such data are not available except for a few common species over an extended temperature range. A method for predicting such data was devised by Criss and Cobble which is known as the "Correspondence Principle." It can be summarised as follows:

"A standard state can be chosen at every temperature such that the partial molar entropies of one class of ions at the temperature are linearly related to the corresponding entropies at some reference temperature".

Choosing the reference temperature as $T^o = 298K$, they proposed the following relationship:

$$S_a^o(T) = a(T) + b(T)S_a^o(298) \tag{3.32}$$

where $a(T)$ and $b(T)$ are empirical parameters and depend on the type of ion being considered and temperature. $S_a^o(T)$ and $S_a^o(298)$ are defined as the "absolute" ionic entropies and are related to "conventional" entropies by:

$$S_a^o(298) = S^o(298) - 20.9n \tag{3.33}$$

where n is the ionic charge. It is important to note that the choice of standard state for "absolute" entropy corresponds to an ionic entropy of $-20.9 \, J \, mol.K^{-1}$ for $H^+$ at 298K.

The average heat capacity between the reference temperature and the operating temperature T can be obtained from the fundamental thermodynamic relationship

$$S_a^o(T) = S_a^o(298) + \int\limits_{298}^{T} \left( C_p^o / T \right) dT \qquad (3.34)$$

$$S_a^o(T) = S_a^o(298) + C_p^o \Big]_{298}^{T} \int\limits_{298}^{T} (1/T)\, dT \qquad (3.35)$$

where $C_p^o\big]298^T$ is the average value over the temperature range 298 to T.

Substituting the Criss-Cobble equation, one obtains

$$a(T) + b(T)\, S_a^o(298) - S_a^o(298) = C_p^o\big]298^T . \ln(T/T^o) \qquad (3.36)$$

which on rearrangement becomes

$$C_P^o\big]_{298}^{o} = \alpha(T) + \beta(T) S_a^o(T^o) \qquad (3.37)$$

$$\text{in which } \alpha(T) = \frac{a(T)}{\ln(T/T^o)} \text{ and } \beta(T) = \frac{(1 - b(T))}{\ln(T/T^o)} \qquad (3.38)$$

Values of $\alpha(T)$ and $\beta(T)$ are given in Appendix A7 of Chapter 2 for various types of ions.

The Criss and Cobble correspondence principle is particularly useful for determining equilibrium constants of components in solution at elevated temperatures. For a reaction at equilibrium at a temperature T

$$\Delta G_R^o(T) = \Delta H_R^o(T) - T\Delta S_R^o(T) \qquad (3.39)$$

$$\Delta H_R^o(T) = \Delta H_R^o(298) + \int\limits_{298}^{T} \Delta C_P^o(T) dT \qquad (3.40)$$

$$\Delta S_R^o(T) = \Delta S_R^o(298) + \int\limits_{298}^{T} \left( \Delta C_P^o(T) / T \right) dT \qquad (3.41)$$

Replacing $\Delta C_P^o(T)$ by an average heat capacity and combining the above equations:

$$\Delta G_R^o(T) = \Delta G_R^o(T^o) + (T - T^o)\Delta C_p]_{298}^T - T(\ln T / T^o)\,\Delta C_p]_{298}^T$$

$$(3.42)$$

For $H_2O$ and other soluble species as well as solids and gases, $C_p^o$ data is often available in the form

$$C_p^o(T) = a + bT + c/T^2 + d.T^2 \qquad (3.43)$$

where a, b, c and d are tabulated parameters. An average heat capacity can be obtained from

$$C_P^o]_{298}^T = \frac{1}{T - 298} \int_{298^o}^T C_P^o dT \qquad (3.44)$$

The following steps are recommended in order to use the Correspondence Principle to calculate thermodynamic data such as $\Delta G^o$ and K at elevated temperatures.

1. Use tables of thermodynamic data to calculate $\Delta G^o$, $\Delta H^o$ and $\Delta S^o$ for the reaction at 298K.
2. Calculate $S_{abs}^o$ at 298K for each species from,

$$S_{abs}^o = S^o - 20.9z \left(Jmol^{-1}K^{-1}\right) \qquad (3.45)$$

where z is the ionic charge.

3. Calculate $C_p]_{298}^T$ for each species using tabulated values of "$\alpha(T)$" and "$\beta(T)$" for the temperature T,

$$C_p]_{298}^T = \alpha(T) + \beta(T).S_{abs}^o \qquad (3.46)$$

4. Calculate $\Delta C_p$ for the reaction using the above $C_p$ values.
5. Calculate $\Delta G_T^o$ for the reaction from,

$$\Delta G_T^o = \Delta G_{298}^o - (T - 298).\Delta S_{298}^o + (T - 298 + T.\ln(298 / T)).\Delta C_p]_{298}^T$$

$$(3.47)$$

Note.

Note that for species other than ions, you can calculate the values of $\Delta C_p]_{298}^T$ by using the empirical equation above.

For the reaction

$$Fe^{3+}(aq) + H_2O(l) \rightleftharpoons FeOH^{2+}(aq) + H^+(aq)$$

the value of log K at 25°C is −2.19.

Determine log K for this reaction at 100°C.

Solution:

$$\Delta G_r^o(298) = 2.3RT \log K$$

$$= (2.3)(8.314)(298)(2.19)$$
$$= 12480 \text{ J mol}^{-1}$$

Using entropy data at 25°C from the Appendix in Chapter 2,

$$\Delta S_r^o(298) = 142 + 0 + 316{-}70$$
$$= 104 \text{ J (mol.K)}^{-1}$$

Using the data in the Appendix in Chapter 2 and the Criss-Cobble equation in step 3 for the three ionic species:

$$H^+(aq): C_p \ (298{-}313K) = 130 \text{ J (mol.K)}^{-1}$$

$$Fe^{3+}(aq): C_p \ (298{-}313K) = 192{-}0.55\{-316{-}3(20.9)\}$$

$$= 401 \text{ J (mol.K)}^{-1}$$

$$FeOH^{2+}(aq): C_p \ (298{-}313K) = 192{-}0.55\{-142{-}2(20.9)\}$$

$$= 294 \text{ J (mol.K)}^{-1}$$

$$H_2O(l): C_p \ (298{-}313K) = 76 \text{ J (mol.K)}^{-1}$$

For the reaction at 373K,

$$\Delta C_p^o(373) = 294 + 130{-}401{-}76$$
$$= -53 \text{ J (mol.K)}^{-1}$$

Using the equation in Step 5,

$$\Delta G_R^o \ (373 \ K) = \Delta G_R^o \ (298) - (373 - 298) \ \Delta S_r^o(298)$$
$$+ (373 - 298) \ \Delta C_p^o \ (373) - 373 \ln(373/298) \ \Delta C_p^o \ (373)$$

$$= 12480{-}7800{-}3975 + 4438 = 5143 \text{ J mol}^{-1}.$$

$$\text{Log } K = -5143/(2.3 \times 8.314 \times 373) = -0.72$$

that is Iron(III) is more strongly hydrolysed at higher temperatures.

## 3.4 Redox equilibria

In most hydrometallurgical processes, changes of oxidation state occur in some of the elements taking part and manipulation of oxidation states is a powerful tool at the disposal of hydrometallurgists.

This section is largely revision of the most important concepts and applications of electrochemical potentials as they apply to the equilibria of reactions taking part during the processing of minerals and metals. The important extension into the rates (or kinetics) of redox reactions will be dealt with in Chapter 5.

### 3.4.1 Electrochemical potentials

The reaction

$$Zn + CuSO_4(aq) = Cu + ZnSO_4(aq) \qquad (3.48)$$

is known to proceed spontaneously.

It can (theoretically and practically) be separated into two half reactions

$$Zn = Zn^{2+} + 2e \qquad (3.49)$$

$$Cu^{2+} + 2e = Cu \qquad (3.50)$$

Consider a cell composed of these two half-reactions as shown in Fig. 3.3.

For passage of current I for a time, t,

$$\text{Work done in external circuit } W = I.t.V = Q.V \qquad (3.51)$$

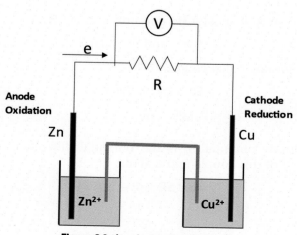

**Figure 3.3** An electrochemical cell.

For conversion of 1 mole of Cu $Q = 2q_e.N$

$$= 2F(\text{Faraday } 96487 \text{ coulomb mol}^{-1}) \qquad (3.52)$$

Work done $\text{mol}^{-1}$ $W = nFV$ in which $n$ = stoichiometric no. of electrons.

As R increases, V will increase and in the limit $R \rightarrow \infty$, V approaches the maximum thermodynamic cell voltage, $\Delta E$.

Under these conditions, max $W = -\Delta G = nF\Delta E \qquad (3.53)$

## 3.4.2 Single electrode potentials

In the case of the above cell, a voltage (or potential difference) of 1.1 V would be measured by the voltmeter if the solutions contained copper and zinc ions of concentrations equivalent to unit activity. The copper electrode would be positive with respect to the zinc electrode.

The cell voltage can be considered to be composed of a difference between the potential of the copper electrode and that of the zinc electrode,

$$\text{i.e. } \Delta E = V = E_{cu} - E_{zn} \qquad (3.54)$$

Note that it is only possible to measure the potential difference (or voltage) between two electrodes.

A scale of relative potentials can be devised if all potentials are measured relative to one electrode. For various reasons this standard reference electrode has been chosen as the standard hydrogen electrode (SHE)

$$2H^+ + 2e = H_2 \qquad (3.55)$$

for which, at unit activity $H^+$ and unit fugacity $H_2$,

$$\Delta G^o = 0 \text{ and } E^o = 0.0V.$$

Thus for any redox couple,

$$Ox + ne = \text{Red or}$$

$$M^{n+} + ne = M \qquad (3.56)$$

the electrode potential(E) is the potential difference of a cell composed of this couple and a SHE.

$$\text{i.e. } \Delta E = E - E_H^o \qquad (3.57)$$

and a positive $\Delta E$ (or negative $\Delta G$) implies that the reaction

$$Ox + H_2 \rightarrow Red + 2H^+$$

proceeds spontaneously.

Note that all reactions are written as reductions, that is, with electrons on the left-hand side. A positive E for a couple means that hydrogen is a stronger reducing agent than the reduced form of the couple.

For example, the reaction

$$2Cu(s) + O_2(g) + 4H^+(aq) = 2Cu^{2+}(aq) + 2H_2O \qquad (3.58)$$

is made up of the difference of the half reaction

$$O_2(g) + 4H^+(aq) + 4e = 2H_2O \qquad (3.59)$$

and twice that of the half reaction

$$Cu^{2+}(aq) + 2e = Cu(s) \qquad (3.60)$$

with a hypothetical cell voltage (under standard conditions) given by

$$\begin{aligned} \Delta E^o = V = E^o_{O_2} - E^o_{Cu} \\ = 1.229 - 0.340 = 0.889 \text{ V} \end{aligned} \qquad (3.61)$$

Thus, under these conditions,

$$\begin{aligned} \Delta G^o = -nF\Delta E^o \\ = -4.\ 96500.\ 0.889 \\ = -343.2 \text{ kJ(mol O}_2)^{-1} \end{aligned} \qquad (3.62)$$

The large negative value for $\Delta G^o$ implies a spontaneous reaction as written.

Table 3.2 gives a selection of the $E^o$ values of some couples of importance in hydrometallurgical processes. More extensive tabulations are available in several published compilations.

Thus, electrode potentials can simply be added or subtracted to give a potential difference for an overall reaction, but this is not true for another half-reaction. In this case, the free energies, that is volt equivalents (nE) of the two half reactions must be added or subtracted to give the free energies of the third half-reaction that can be converted back to a potential.

Thus, for example, one can use the potentials for the reactions

$$Cu^{2+} + 2e = Cu \text{ 0.34 V or 0.68 V equivalents} \qquad (3.63)$$

**Table 3.2 Some relevant standard reduction potentials (Milazzo et al., 1978; Bratsch, 1989; Bard et al., 1983).**

| Half reaction | $E^o$, volts |
|---|---|
| $Mn^{2+} + 2e = Mn$ | −1.18 |
| $Zn^{2+} + 2e = Zn$ | −0.76 |
| $Au(CN)_2^- + e = Au + 2CN^-$ | −0.57 |
| $Fe^{2+} + 2e = Fe$ | −0.44 |
| $Cd^{2+} + 2e = Cd$ | −0.40 |
| $Ag(CN)_2^- + e = Ag + 2CN^-$ | −0.37 |
| $PbSO_4 + 2e = Pb + SO_4^{2-}$ | −0.35 |
| $Co^{2+} + 2e = Co$ | −0.28 |
| $Ni^{2+} + 2e = Ni$ | −0.26 |
| $Pb^{2+} + 2e = Pb$ | −0.13 |
| **$2H^+ + 2e = H_2$** | **0** |
| $SO_4^{2-} + 4H^+ + 2e = H_2SO_3 + H_2O$ | 0.17 |
| $AgCl + e = Ag + Cl^-$ | 0.222 |
| $Cu^{2+} + 2e = Cu$ | 0.34 |
| $Fe^{2+} + S + 2e = FeS_2$ | 0.34 |
| $Fe^{2+} + 2SO_4^{2-} + 16H^+ + 14e = FeS_2 + 8H_2O$ | 0.36 |
| $SO_4^{2-} + 8H^+ + 6e = S + 4H_2O$ | 0.36 |
| $UO_2^{2+} + 2e = UO_2$ | 0.41 |
| $Cu^{2+} + Fe^{2+} + 2S = CuFeS_2$ | 0.41 |
| $Cu^+ + e = Cu$ | 0.52 |
| $Fe^{3+} + e = Fe^{2+}$ | 0.77 |
| $Ag^+ + e = Ag$ | 0.80 |
| $O_2 + 4H^+ + 4e = 2H_2O$ | 1.23 |
| $MnO_2 + 4H^+ + 2e = Mn^{2+} + 2H_2O$ | 1.33 |
| $Cl_2 + 2e = 2Cl^-$ | 1.36 |
| $PbO_2 + SO_4^{2-} + 4H^+ + 2e = PbSO_4 + 2H_2O$ | 1.70 |
| $H_2O_2 + 2H^+ + 2e = 2H_2O$ | 1.76 |
| $NiOOH + 3H^+ + e = Ni^{2+} + 2H_2O$ | 1.80 |

Note that the electrode potential is an intensive property (independent of the quantity of material), whereas the free energy is an extensive property. Bold represents the central point separating two classes.

$$Cu^+ + e = Cu \ 0.52 \text{ V or } 0.52 \text{ V equivalents} \qquad (3.64)$$

to estimate that for the reaction

$$Cu^{2+} + e = Cu^+$$

by subtracting Eq. (3.64) from Eq. (3.63) to give 0.16V and **not** $-0.18V$.

> What are the half reactions taking place in a lead-acid battery used in automobiles? What is the theoretical cell voltage? How do we get 12V in an actual battery?

### 3.4.3 Reference electrodes

The hydrogen electrode is, in most cases, not convenient for use as a reference electrode and therefore other electrodes with suitable properties are used, the most common of which is the silver/silver chloride electrode, which is based on the reaction,

$$AgCl(s) + e = Ag(s) + Cl^-(aq) \tag{3.65}$$

which has a potential of 0.199 V versus SHE at 25°C with saturated KCl electrolyte. The electrode is simply a partially anodised silver wire immersed in a solution of potassium chloride.

### 3.4.4 Nonstandard electrode potentials

Most reactions are carried out under conditions in which the activities of the species taking part are not unity and one must be able to quantitatively account for deviations from the standard states. For a general redox reaction,

$$aA + bB = cC + dD \tag{3.66}$$

$$\Delta G = \Delta G^o + RT \ln\{a_C^c \cdot a_D^d \,/\, a_A^a \cdot a_B^b\} \tag{3.67}$$

where a's are the activities of the species.

Similarly, because of the relationship between $\Delta G$ and E ($\Delta G = -nFE$),

$$E = E^o - (RT\,/\,nF) \cdot \ln\{a_C^c \cdot a_D^d \,/\, a_A^a \cdot a_B^b\} \tag{3.68}$$

This is the well-known Nernst equation which can be applied equally to half-reactions, such as

$$Ox + ne = Red$$

$$E = E^o - (RT\,/\,nF) \cdot \ln\{a_{red}\,/\,a_{ox}\} \tag{3.69}$$

For a general half-reaction,

$$aA + bB + mH^+ + ne = cC + dD \qquad (3.70)$$

where A and C are the oxidised and reduced species of interest and B and D are auxiliary species such as water or complexing agents.

$$E = E^o$$
$$- 2.303m(RT/nF) . pH - 2.303(RT/nF) . \log\{a_C^c . a_D^d / a_A^a . a_B^b\} \qquad (3.71)$$

Try and remember that $2.303RT \, F^{-1} = 0.0591$ V at 25°C and that concentrations are often used in place of activities in dilute solutions.

For example, the potential of the couple

$$Au(CN)_2^- + e = Au + 2CN^- \qquad (3.72)$$

for $10^{-3}$mol $L^{-1}$ $CN^-$ and $10^{-5}$mol $L^{-1}$ $Au(CN)_2^-$ is given by

$$E = E^o - 2.303(RT/F) . \log\{[CN^-]^2 / [Au(CN)_2^-]\} \qquad (3.73)$$

where concentrations are used instead of activities with little error at these low concentrations.

Substituting the values for $E^0$ and the concentrations,

$$E = -0.57 - 0.0592 \log(10^{-3})^2/10^{-5}$$
$$= -0.57 + 0.0592 = -0.51 \text{ V} \qquad (3.74)$$

Note that this potential is very much lower than that for the reaction

$$Au^+ + e = Au \qquad\qquad E^o = 1.68 \text{ V}$$

and this shows the effect of strong complexation of the aurous ion by the cyanide ion.

## 3.4.5 Effect of complexation on redox potentials

The formation of complexes between a metal ion and a ligand can have a significant effect on the value of the standard reduction potential.

Thus, consider the example,

$$Ni^{2+} + 2e = Ni \qquad\qquad E_1^o = -0.26 \text{ V} \qquad (3.75)$$

In the presence of complexing ligands such as $Cl^-$ or $NH_3$, this potential will change. The effect of complexation can be viewed as a reduction in the activity of the free $Ni^{2+}$ ion which will make it "more difficult" for it to be reduced to metallic nickel, that is one will require a more negative potential. This effect can be quantitatively accounted for as follows:

Thus, for the hexamine complex,

$$Ni^{+2} + 6NH_3 = Ni(NH_3)_6^{2+} \quad \log \beta_6 = 8. \tag{3.76}$$

from which

$$\left(Ni^{+2}\right) = \left(Ni(NH_3)_6^{2+}\right) / \left\{ (NH_3)^6 \cdot \beta_2 \right\} \tag{3.77}$$

The Nernst equation for Eq. (3.73) can be written as

$$E_1 = E_1^o - (RT / 2F) \cdot \ln\left\{ 1 / \left(Ni^{+2}\right) \right\} \tag{3.78}$$

and, substituting for $[Ni^{+2}]$ from Eq. (3.77),

$$E_1 = E_1^o - (RT / 2F) \cdot \ln \beta_2 - (RT / 2F) \cdot \ln\left\{ (NH_3)^6 / \left(Ni(NH_3)_6^{2+}\right) \right\} \tag{3.79}$$

For the reduction of the complexed ion,

$$Ni(NH_3)_6^{2+} + 2e = Ni + 6NH_3 \tag{3.80}$$

the Nernst equation is,

$$E_6 = E_6^o - (RT / 2F) \cdot \ln\left\{ (NH_3)^6 / \left(Ni(NH_3)_6^{2+}\right) \right. \tag{3.81}$$

Comparison of Eqs (3.79) and (3.81) shows that,

$$E_6^o = E_1^o - (RT/2F) . \ln \beta_6$$
$$= -0.26 - (0.0591/2) . \log 10^{8.1} \tag{3.82}$$
$$= -0.50V$$

that is a decrease of about 0.25V in the standard reduction potential due to the formation of the ammine complex. The actual potential will obviously vary with both the concentration of ammonia and the pH, and the easiest approach is to calculate the activity of the free $Ni^{2+}$ ion under the particular conditions and substitute into Eq. (3.78).

Note that the so-called formal potential is the equilibrium potential for a couple under conditions of 1M total concentration of each of the electroactive species. Thus, in the above example, one could say that the formal potential of the nickel(II)/Ni couple is -0.50V in a solution of 1M nickel ions and 1M ammonia at 25°C.

## 3.4.6 Diagrams for summarising reduction potentials

There are several methods often used for the depiction of the redox characteristics of an element in its various oxidation states. Thus, Fig. 3.4 shows one such diagram for manganese in acidic solutions in which the highest oxidation state is on the left and the lowest on the right.

The standard reduction potentials for the reduction half-reaction involving any two species joined by a line are shown. The more positive the standard reduction potential, the more readily the species on the left is reduced to the species on the right side of the arrow. Thus, highly positive reduction potentials indicate that the species on the left (such as permanganate) is a powerful oxidising agent. Negative reduction potentials indicate that the species to the right behaves as a reducing agent (such as manganese metal).

An alternative, more useful diagram is shown in Fig. 3.5. In this case, the volt-equivalent $nE^\circ$ (or $-\Delta G^\circ/F$) for the reduction of the element in a given oxidation state to the element in oxidation state zero is plotted against the oxidation state of the element.

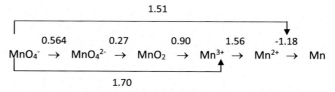

**Figure 3.4** Latimer diagram for manganese in acidic solutions.

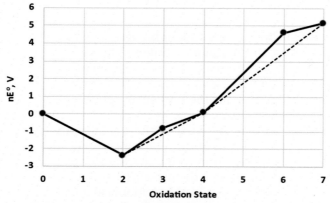

**Figure 3.5** Oxidation state diagram for manganese in acidic solutions.

The following information can generally be derived from an oxidation state diagram:

- Thermodynamic stability is found at the bottom of the diagram. Thus, the lower a species is positioned on the diagram, the greater its thermodynamic stability.

  *$Mn^{2+}$ is the most stable species.*

- A species located **above** the straight line between two oxidation states is unstable with respect to the species at the ends of the line and can undergo disproportionation to the species at the ends of the line.

  *Thus, $MnO_4^{2-}$ will disproportionate to $MnO_4^-$ and $MnO_2$ as shown by the dotted line and $Mn^{3+}$ will disproportionate to $Mn^{2+}$ and $MnO_2$. Note that these species will therefore normally not appear on an $E_H/pH$ diagram (see below). The reader can confirm that, for example the disproportionation reaction*

$$2Mn^{3+} + 2H_2O = Mn^{2+} + MnO_2 + 4H^+ \qquad (3.83)$$

  *is spontaneous, that is, has $\Delta G^o < 0$. Remember that for a reaction $\Delta G^o = -nF\Delta E^o$.*

- The opposite is true of an oxidation state **below** the line joining two oxidation states.

  *$MnO_2$ does not disproportionate to $Mn^{2+}$ and $MnO_4^-$.*

- Any species located on the **upper right** side of the diagram will be a strong oxidising agent.

  *$MnO_4^-$ is a strong oxidiser.*

- Any species located on the **upper left** side of the diagram will be a reducing agent.

  *Manganese metal is a strong reducing agent.*

- These diagrams describe the thermodynamic stability of the various species.

  Although a given species might be thermodynamically unstable toward reduction, the kinetics of such a reaction might be very slow.

  *Although it is thermodynamically favourable for permanganate ion to be reduced to $Mn^{2+}$ by reduction with water, the reaction is slow except in the presence of a catalyst. Thus, solutions of permanganate can be stored and used in the laboratory.*

- The information obtained from an oxidation state diagram is for species under standard conditions (pH=0 for acidic solution and pH=14 for basic solution). Changes in pH may change the relative stabilities of the species. The potential of any

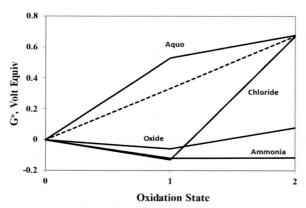

**Figure 3.6** Oxidation state diagram for copper.

process involving the hydrogen ion will change with pH because the concentration of this species is changing.

*Under basic conditions aqueous Mn²⁺ does not exist. Instead, insoluble Mn(OH)₂ forms.*

The oxidation state diagram for copper is shown in Fig. 3.6 for various solutions and also for the solid oxides. In the case of copper, there are only two soluble oxidation states with copper(II) being the most stable.

Note:

- The slope of the line joining two oxidation states is the $E°$ for that couple.
- Again, because the point for the $Cu^+$ aquaion lies above the dotted line, it is unstable and will disproportionate to Cu(s) and $Cu^{2+}$.
- On the other hand, $Cu_2O$ is stable
- In the presence of ligands that form complexes with $Cu^+$, copper(I) is stable, and this is the case for ammonia and chloride. This has important consequences for the hydrometallurgy of copper.

The diagram for sulphur is shown in Fig. 3.7.

In this case, the potentials for the many oxidation states vary with pH except for $S°$, thus two cases are shown for pH 0 and pH 10. One can conclude that neither sulphide ion (or hydrogen sulphide) and elemental sulphur are useful reducing agents while sulphate (oxidation state 6) is thermodynamically an oxidising agent. However, the extremely low reactivity of the sulphate or bisulphate ion to reduction is such that it is to all intents unreactive. However, most of the intermediate oxidation states are reactive and many of the oxidation states are unstable with respect to

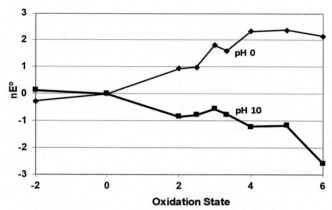

**Figure 3.7** Oxidation state diagram for sulphur.

disproportionation to the end members ($S^o$ or $S^{2-}$ and $SO_4^{2-}$) because they lie above the dotted lines line connecting the end states at both pH values.

What is the oxidation state of S in dithionite, tetrathionate and thiosulphate?
Is tetrathionate stable relative to thiosulfate and dithionite at pH 0 and pH 10? Why?

As a result of the instability of the intermediate oxidation states of sulphur, these species will not appear on a normal Eh/pH diagram (see next section).

## 3.5 Eh/pH diagrams

These diagrams are a convenient means of graphically displaying aqueous solution thermodynamic data as a function of the pH and electrochemical potential. They are invaluable in interpreting and predicting the behaviour of ions in solution and are useful in describing the chemistry of all hydrometallurgical operations.

## 3.5.1 Electrochemical stability of water

The stability of water as a solvent is determined by two equilibria involving on one hand, reduction to hydrogen gas and on the other, oxidation to oxygen gas.

Thus, consider the reaction

$$H^+ + e = 1/2 \, H_2$$

Note that this reaction can also be written as the equivalent reaction

$$H_2O + e = 1/2 \, H_2 + OH^- \tag{3.84}$$

The Nernst equation can be written as

$$E = E^o - \frac{RT}{nF} \ln \frac{\left(p_{H_2}\right)^{1/2}}{(H^+)}$$

$$= E^o - 0.0591 \log\left(p_{H_2}\right)^{1/2} - 0.0591(-\log(H^+))$$

for $T = 298K$ and $n = 1$.

$$= 0 - 0.0591 \log(1)^{1/2} - 0.0591 \log pH = 0 - 0.0591$$

pH for $p_{H_2} = 1$. \hfill (3.85)

A plot of E versus pH for this couple will have a slope of $-0.059$ V and a y-intercept of 0 V as shown as the lower dotted line (a) in Fig. 3.8.

Similarly, for the reaction,

$$O_2 + 4H^+ + 4e = 2H_2O \tag{3.86}$$

$$E = E^o - \frac{RT}{4F} \ln \frac{1}{\left(p_{O_2}\right)(H^+)^4}$$

$$= 1.23 + \frac{0.0591}{4} \log p_{O_2} - \frac{0.0591}{4}[+4 \, pH]$$

$$= 1.23 - 0.0591 \, pH \tag{3.87}$$

A plot of E versus pH for this couple will have a slope of $-0.059$ V and an intercept at pH 0 of 1.23 V as shown as the upper dotted line (b) in Fig. 3.8.

Thus, water is stable at potentials within the region bounded by the two dotted lines and will be oxidised at potentials above the upper line and reduced at potentials below the loer line. However, as we will see in a later chapter, this area of stability can be extended somewhat due to the kinetic inertness of water

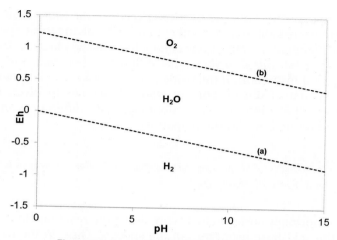

**Figure 3.8** $E_H - pH$ diagram for water at 25°C.

at temperatures most commonly encountered in hydrometallurgical processes. In some cases, it is important to include two extra lines for the reactions

$$O_2 + 2H^+ + 2e = H_2O_2$$

and

$$H_2O_2 + 2H^+ + 2e = 2H_2O$$

because the reduction of oxygen often occurs through peroxide as an intermediate. See the diagram for the gold-cyanide system in Case study 2. for example.

## 3.5.2 The diagram for the zinc — water system

The construction of these diagrams is relatively simple and involves a number of simple steps. The procedure will be illustrated using a relatively simple example of the chemistry of zinc in aqueous solutions. In the absence of complex ligands, the following species of zinc are formed

Solids: $Zn$, $ZnO$, $Zn(OH)_2$

Aqueous species: $Zn^{2+}(aq)$, $Zn(OH)_4^{2-}(aq)$ (or $ZnO_2^{2-}$)

The species $ZnOH^+(aq)$ is also present at certain pH values but will be ignored for the present.

Note that ZnO and $Zn(OH)_2$ differ only by one water molecule and for this reason, the reaction equations involving these species are very similar and either one can be used in the diagram. In the case of most metal oxides, the difference between the free energies of formation of the oxide and the hydroxide is generally close to that of the number of water molecules involved with the result that there are generally only small differences in the position of the lines on the diagrams.

By looking up tables of thermodynamic data, check whether the above is true for the metals Cu, Ni, Zr, Cr(III).

It is normally easier to first write down the reactions involving only the acid-base properties of the species. Thus, in the case of zinc, these are

1(a).     $Zn^{2+} + H_2O = ZnO(s) + 2H^+$ or

1(b).     $Zn^{2+} + 2OH^- = ZnO(s) + H_2O$     (3.88)

and

2.     $ZnO(s) + H_2O = Zn(OH)_4^{2-} + 2H^+$     (3.89)

Note that the two equilibria 1(a) and 1(b) are equivalent and will give rise to the same line on the diagram. It is generally easier to write such equilibria in the form of (a) above as it involves $H^+$ directly. This makes it easier to convert to pH in the expressions to describe the equilibrium.

Thus, using 1(a), the equilibrium involved can be written as

$$K_1 = [ZnO] \cdot [H^+]^2 / [Zn^{2+}] \cdot [H_2O] = [H^+]^2 / [Zn^{2+}]$$     (3.90)

Taking logarithms and rearranging, recognising that $pH = -\log[H^+]$, we obtain the simple relationship,

$$pH_1 = \text{-}0.5 \log K_1 \text{ for } [Zn^{2+}] = 1.$$

In order to plot this relationship, we require a value for $K_1$ which can be obtained from tables or, more generally by calculation using tabulated thermodynamic data such as that in Table 3.3.

Thus,

$$\Delta G_1^\circ = \Delta G^\circ(ZnO) - \Delta G^\circ(Zn^{2+}) - \Delta G^\circ(H_2O)$$
$$= -318.3 - (-147.2) - (-237.18)$$     (3.91)
$$= -2.303 \, RT.\log K_1$$

**Table 3.3 Thermodynamic data for zinc/water system at 298K.**

| Species | $\Delta G^\circ$, kJ mol$^{-1}$ |
|---|---|
| ZnO(s) | −318.3 |
| Zn(OH)$_2$(s) | −461.6 |
| Zn$^{2+}$(aq) | −147.2 |
| Zn(OH)$_4^{2-}$(aq) | −877.4 |
| ZnOH$^+$(aq) | −342.0 |
| OH$^-$(aq) | −157.4 |
| H$_2$O(l) | −237.18 |

from which,

$$\log K_1 = -10.96 \text{ and } pH_1 = -0.5 \log K_1 = 5.48 \qquad (3.92)$$

This can be plotted as a vertical line at pH 5.48 on the diagram.

We can similarly calculate the position for the line describing equilibrium (2)

$$ZnO(s) + 3H_2O = Zn(OH)_4^{2-} + 2H^+ \qquad (3.93)$$

to give

$$pH_2 = -0.5 \log K_2 = 13.4 \text{ for} \left[ Zn(OH)_4^{2-} \right] = 1 \qquad (3.94)$$

that is, a vertical line at pH 13.4.

Derive the above expression for pH$_2$ and calculate logK$_2$ using the data in the above table of thermodynamic data.

We can now derive the equations for the lines, which describe the equilibria for those reactions that involve redox processes that is, those reactions that result in the change of oxidation state of at least one of the elements. In this case, the simplest of these involves the equilibrium between zinc metal and the aquo-zinc ion, which can be described by the Nernst equation. Thus, for the reaction

3.
$$Zn^{2+} + 2e = Zn \qquad (3.95)$$

$$E_3 = E_3^o - RT/nF. \ln(\{Zn\}/\{Zn^{2+}\})$$

$$= -0.763 - 0.0591. \log(1/\{Zn^{2+}\}) = -0.763 \text{ V}$$

$$\text{for}\{Zn^{2+}\} = 1 \text{ and T} = 25°C$$

$$(3.96)$$

This can be plotted as a horizontal line at $E = -0.763$ V on the diagram.

Similarly, the equilibrium between zinc metal and zinc oxide can be expressed by the equation

4.        $ZnO(s) + 2H^+ + 2e = Zn + H_2O$       (3.97)

with the corresponding Nernst expression

$$E_4 = E_4^o - RT/2F . \ln(1/\{H\}^2)$$

$$= -0.439 - 0.0591 \text{ pH at } 25°C$$

$$(3.98)$$

Note that this expression involves the pH and can be plotted on the diagram as a line of intercept $-0.439$v at pH 0 and a negative slope of 0.0591V per pH unit as shown.

> Using the thermodynamic data in the table above, calculate the value of $E_4^o$ used in the above expression.

Finally, the equilibrium between the zincate ion and zinc metal can be expressed by the equation

5.        $Zn(OH)_{4}^{2-} + 4H^+ + 2e = Zn + 4H_2O$      (3.99)

for which the corresponding Nernst expression is,

$$E_5 = E_5^o - 4. 2.303RT/2F. \text{ pH} = 0.441 - 0.1182 \text{ pH} \quad (3.100)$$

which can be plotted as a line of intercept 0.441 V and slope $-0.1182$ V per pH unit.

In addition to these lines, all diagrams show two additional (generally dotted) lines, which describe the equilibria involved in the stability of water as derived in Section 3.5.1.

The Eh/pH diagram shown in Fig. 3.9 now shows all of these lines.

Note that we have eliminated some sections of these lines, which describe equilibria that are not relevant in certain areas. Thus, for example, line (3) cannot be extended beyond a pH value

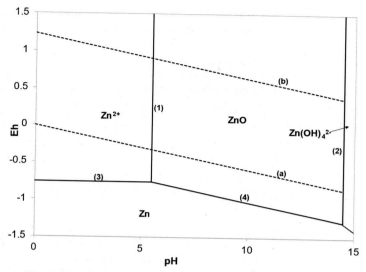

**Figure 3.9** Complete $E_H - pH$ diagram for the zinc-water system.

of 5.48 because the $Zn^{2+}$ ion is not stable at higher pH values. Similarly, the vertical line (1), which separates $Zn^{2+}$ from ZnO cannot extend below $-0.763$ V because $Zn^{2+}$ is reduced to Zn metal at potentials below this value.

The areas of stability of the various species can now be identified on the diagram. Thus, the $Zn^{2+}$ ion is only stable at a concentration equivalent to unit activity in the rectangular area bounded by lines (1) and (3) on the left-hand side of the diagram.

The Appendix to this chapter has a number of such diagrams, which you will need to refer to in various other chapters and you will also be required in order to answer the practice problems for this chapter. The most extensive compilation of such diagrams for the water/element systems is available (Pourbaix, 1966).

### 3.5.3 General procedure for constructing diagrams

The above procedure in constructing such diagrams can be generalised into the following steps:

- List the various species formed by the element of interest.
- Write down balanced chemical equations for the equilibria involving these species.
- Establish the relevant concentrations of the various species to be used in the calculations of the equilibria expressions.
- Derive expressions, which mathematically describe the equilibria.

- Substitute the relevant concentrations in these expressions.
- Plot each expression as a line on the diagram.
- Delete sections of the lines, which are not applicable to the pH or $E_h$ region.
- Label the areas of stability of the species.

This procedure can be considerably simplified by using the following general formula for the line describing the equilibrium

$$aA + bB + mH^+ + ne = cC + dD \qquad (3.101)$$

at 25°C,

$$E = E^o - 0.0591/n.\log \frac{\{C\}^c \{D\}^d}{\{A\}^a \{B\}^b} - 0.0591(m/n)\,pH \qquad (3.102)$$

which, for unit activity of the reacting species, gives

$$E = E^o - 0.0591(m/n)\,pH \qquad (3.103)$$

or a straight line of slope $-0.0591$ (m/n) and an intercept at pH=0 of $E^o$.

**Note:**

1. $E^o$ ($=\Delta G^o$) can often be obtained by simply looking up the value in tables of electrochemical potentials.
2. For m = 0, E = $E^o$ i.e. a horizontal line at E = $E^o$.
3. For the case n = 0, i.e. not an electrochemical process, the above equation cannot be used because m/n is infinite and the following equation can be used.

$$pH = \frac{-\Delta G^o}{2.303mRT} - \frac{1}{m}\log\frac{\{C\}^c \{D\}^d}{\{A\}^a \{B\}^b} \qquad (3.104)$$

4. A line describes the equilibrium between two species while at a point of intersection of three lines, three species are in equilibrium under the specified conditions of activities and temperature. This is a rapid means of checking the correctness of the lines and also of drawing the third line from the intersection of two lines.
5. The soluble species concentrations should be specified in a diagram. For example, a diagram with soluble gold in a cyanide system would not be useful at 1M gold concentration.
6. There are several software packages now available, which can construct such diagrams and often contain thermodynamic databases so that one simply is required to specify the elements of interest and their activities if soluble.

## 3.5.4 Interpretation of diagrams

Several general features of these diagrams should be pointed out:

- The oxidation state of the metal (or other central species) increases as one moves vertically in the diagram.
  In the case of the zinc diagram, the area of stability of Zn(II) lies above that of Zn(0).
- The extent of hydrolysis increases from left to right.
  In the above case, the extent of hydrolysis increases from $Zn^{2+}$ to $Zn(OH)_2$ to $Zn(OH)_4^{2-}$ as we move from left to right.
- Soluble species are generally present on the left and sometimes (for amphoteric metals) on the right of the diagram.
- The area of stability of water is between the two dotted lines (a) and (b). The consequence of this is that species with an area of stability below (a) will be unstable in aqueous solutions reducing water to hydrogen gas.
  Thus, because the line separating $Zn^{2+}$ from Zn(s) is below the line (a), the reaction

$$Zn(s) + 2H^+ = Zn^{2+} + H_2(g)$$

is thermodynamically favourable at pH values below about 5.5 while the reaction

$$Zn(s) + H_2O = Zn(OH)_2(s) + H_2(g)$$

should occur at higher pH values.

> If this is true, why is steel sheeting galvanised that is, covered with a layer of zinc metal, to minimise atmospheric corrosion?

- Similarly, the oxidised forms of species with an area of stability which lies above line (b) will oxidise water to oxygen.

## 3.5.5 Some other examples.

A simplified diagram of the $U - H_2O$ system is shown in Fig. 3.10.

Several features of this diagram reinforce the general features described above.

- The oxidation state increases from 0 to +3 to +4 to +6 for soluble uranium species and from 0 to +4 to +6 to +8 for insoluble species as one moves vertically in the diagram.
- Uranium metal is unstable in aqueous solutions over the whole pH range because its area of stability is outside that

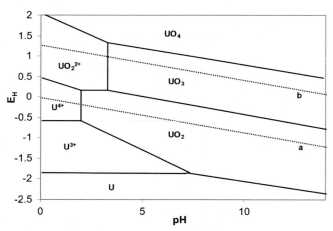

**Figure 3.10** The $E_H$/pH diagram for the uranium - water system (0.01M soluble species).

for water i.e., the lines separating uranium metal from $U^{3+}$ and from $UO_2$ lie well below line (a) i.e., uranium will react with water with the evolution of hydrogen.

- The stable species in aerated aqueous solutions are the uranyl ion, $UO_2^{2+}$ at pH values below about 3 and the oxide, $UO_3$ at higher pH values. Uranium can be precipitated as this oxide by simply increasing the pH of a uranyl solution. This is generally the final step in the extraction of uranium from its ores.
- The +8 oxidation state, represented by the oxide $UO_4$, is unstable in aqueous solutions and will oxidise water to oxygen because the line defining its stability lies above that (b) for the stability of water. It is therefore rarely encountered in hydrometallurgy.

Is the ion $U^{3+}$ stable in acidic aqueous solutions? Why?

Refer to the diagram for sulphur in Fig. 3.11. This figure illustrates another characteristic of these diagrams in that only 3 oxidation states appear on it S(−2), S(0) and S(+6). The other oxidation states such as thiosulfate ($S_2O_3^{2-}$) and the other polythionates and sulphite do not appear because they are not thermodynamically stable with respect to the above three as we saw in the oxidation state diagram in Fig. 3.7. These thermodynamically metastable species are often important and their areas of metastability can also be drawn on the diagram if necessary but only by removing the data for sulphate.

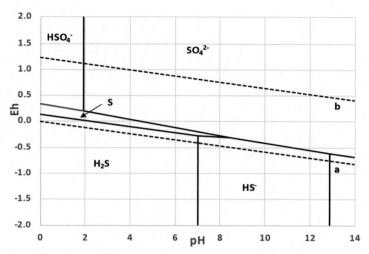

**Figure 3.11** The $E_H$/pH diagram for the sulphur—water system (1M soluble species).

Note the relatively small area of stability of elemental sulphur, which is unstable in aqueous solutions in the presence of oxygen because the lines for its oxidation to sulphate lie below that of the reduction of oxygen, that is line 'b'.

These difficulties can be further illustrated by reference to the zinc-sulphur-water system. Thus, Fig. 3.12 shows the diagram for this system that includes sulphate while Fig. 3.13 shows the same system in which sulphate is excluded. Note that now we have two species (Zn and S) each with its own diagram and the two can be superimposed but the resultant diagram is generally difficult to read and interpret.

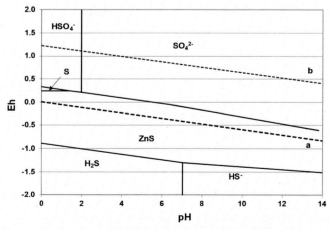

**Figure 3.12** The $E_H$/pH diagram for the zinc sulphide-sulphur-water system (1M soluble species) including sulphate.

Thus, Fig. 3.12 is the S diagram that shows the predominance of sulphate with the area of stability of elemental sulphur being only a small area at low pH. Thus, one would predict from this diagram that oxidation of ZnS by oxygen or iron(III) ($E^o$ =0.75V) will result in sulphate as the product. However, experience shows that elemental sulphur is the product of the oxidative dissolution of ZnS. Redrawing the diagram but excluding sulphate as a species results in the diagram in Fig. 3.13 that shows sulphur as the predominant species. Also shown are the areas of stability of the zinc oxidation products.

The effect of the concentrations of the species can be seen, in this case, in Fig. 3.14.

Thus, the reaction

$$ZnS + H^+ = Zn^{2+} + H_2S \tag{3.105}$$

is known to occur in concentrated acid solutions but will not appear on the diagram at high zinc ion concentrations due the limited solubility of ZnS. However, at low zinc concentrations, a vertical line shown at a pH of about 1 in Figs 3.15 and 3.16 defines this equilibrium for a zinc concentration of $10^{-4}$ M.

It is possible to draw diagrams in which complexed metal ions are included. Thus, the conventional diagram for the copper-water system is shown in Fig. 3.15 for the aqua-ions. The diagram shows that copper(II) is the only soluble species because copper(I) ions are not stable in aqueous solutions in the absence of a complexing ligand such as chloride, ammonia or cyanide. The diagram

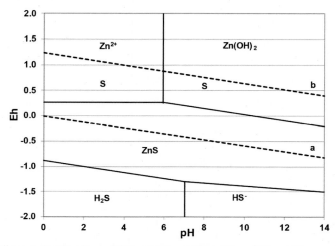

**Figure 3.13** The $E_H$/pH diagram for the zinc sulphide-sulphur-water system (1M soluble species) excluding sulphate.

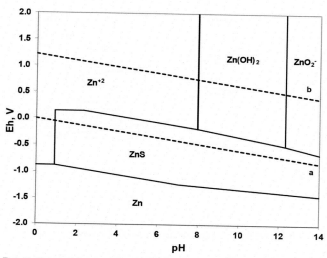

**Figure 3.14** The $E_H$/pH diagram for the zinc sulphide-sulphur-water system ($10^{-4}$M soluble species).

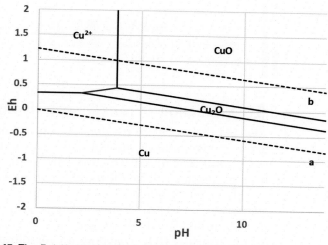

**Figure 3.15** The $E_H$/pH diagram for the copper - water system (1 M soluble species).

can be redrawn to include ammonia as a ligand in species such as $Cu(NH_3)_4^{2+}$ and $Cu(NH_3)_2^+$ as shown in Fig. 3.16.

The area of stability of CuO (or $Cu(OH)_2$) is now partially replaced by copper(II) tetra-amine and that of $Cu_2O$ by copper(I) diamine. Thus, in ammoniacal solutions, copper metal can be oxidised to $Cu^{2+}$ ions at low pH and also to initially copper(I) amine and subsequently to copper(II) amine at pH values between about 8 and 11.

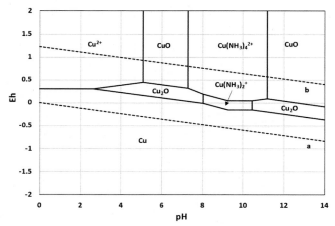

**Figure 3.16** The $E_H$/pH diagram for the copper- ammonia-water system (1 M $NH_3$, 0.1M Cu(II)+Cu(I)).

Why is there a discontinuity at pH 9 in the lines defining the area of stability of $Cu(NH_3)_2^+$ ? Can hydrogen gas reduce copper ions in an ammoniacal solution to copper metal? Why?

### 3.5.6 Prediction of reactions in hydrometallurgy

$E_H$-pH diagrams are invaluable in predicting which reactions are thermodynamically possible and also for defining the approximate conditions under which reactions are likely to take place. These properties will be illustrated by use of the above example of the U-water system.

Uranium occurs in its ores in the form of many minerals, the most important of which are generally the oxides uraninite, $UO_2$ and the trioxide, $UO_3$. Leaching (or dissolution) of the uranium from these minerals requires that one move into the areas of stability of the soluble species in the left-hand region of the diagram shown in Fig. 3.10. Thus, one can dissolve uranium from the trioxide by a simple adjustment of the pH to values below about 3 with formation of the soluble uranyl ion, $UO_2^{2+}$. This is a simple acid leach process which is conventionally carried out using dilute sulphuric acid,

$$UO_3 + 2H^+ = UO_2^{2+} + H_2O \qquad (3.106)$$

In the case of uraninite, there are several possible routes to soluble species.

**(a)** Acid dissolution at pH values below about 2,

$$UO_2 + 4H^+ = U^{4+} + 2H_2O \qquad (3.107)$$

This corresponds to a move in a horizontal direction from right to left to cross the vertical line at pH 2 in the diagram.

**(b)** Movement in a vertical direction to cross the horizontal line at $E_H = 0.2$ V at pH values below about 3.5. Vertical movement implies oxidation

$$UO_2 = UO_2^{2+} + 2e \qquad (3.108)$$

It should be possible to use dissolved oxygen from the air to act as the acceptor for these electrons, that is, the oxidising agent,

$$O_2 + 4H^+ + 4e = 2H_2O \qquad (3.109)$$

because the line (b) lies above the horizontal line at $E_H = 0.2$ V.

> What conditions of pH and $E_H$ are required to thermodynamically ensure that uranium will leach from $UO_2$ by this method? Using the data in Table 3.2, suggest possible chemical reagents that could be used to accomplish this leaching reaction.

The most appropriate oxidant will depend on a number of other considerations, one of the most important of which is the cost.

**(c)** Movement to lower potentials with the formation of the $U^{3+}$ ion. While theoretically possible this option has not been shown to be a practical proposition.

> Find out which oxidising agents are used in U plants in Canada, Namibia and Australia.

The effect of complexing of the uranyl ion by carbonate ions on the U diagram is shown in the diagram in the Appendix.

### 3.5.7 Limitations in the use of $E_H$—pH diagrams.

It should be pointed out that although these diagrams are invaluable in assessing and predicting what processes are *possible* from a thermodynamic point of view, any practical application of a possible reaction will necessarily require that the rate of the reaction is such that it can be made to occur in the time available for processing. Thus, for example, although option (b) above is thermodynamically possible, the rate is very slow except in very concentrated acids at high temperatures. It is therefore of no real practical value.

Similarly, option (c) is of very limited practical value because of the absence of suitable low-cost reductants capable of carrying out the reaction at a reasonable rate.

Option (b) is applied in practice with operating uranium leach plants using oxidants such as ferric ions, sodium chlorate, manganese dioxide or Caro's acid in sulphuric acid solutions at a pH value of about 2. In this case, oxygen is not suitable as an oxidant because, as in many reactions of oxygen, the rate is too slow under normal conditions of temperature and pressure.

Reference to the zinc-water diagram above shows that it should not be possible to electrodeposit zinc from aqueous solutions. This follows from the fact that the line for the reduction of zinc ions lies below line (a), that is, the preferred reaction at the cathode in an electrowinning cell will be

$$2H^+ + 2e = H_2 \tag{3.110}$$

and not the reaction

$$Zn^{2+} + 2e = Zn \tag{3.111}$$

which requires a more negative potential. The fact that zinc is electrowon on a large scale from acidic solutions is due to the fact that the kinetics of the former reaction are very slow on a zinc metal surface, that is the reaction has a high **overpotential**. It is for the same reason that zinc can be used as the anode material having a reasonable shelf-life in common alkaline batteries.

> What is the cell reaction which occurs in an alkaline battery? Estimate the cell voltage by referring to the $E_H$-pH diagrams for zinc-water (above) and manganese-water (Appendix).

**Remember: $E_H$-pH diagrams summarise what reactions are theoretically possible, but the kinetics of the reactions**

**determine what is practically probable.** Unfortunately, there are as yet no simple methods of predicting the rates of most chemical reactions and we have to resort to actual experimentation in order to select the most appropriate reagents and reaction conditions.

## 3.6 Thermodynamic software packages

There are a number of software packages of varying complexity and value available which can rapidly carry out various thermodynamic and equilibrium calculations of the types covered in this course. Many of these packages have their own thermodynamic data bases so that it is often not necessary to look up thermodynamic tables for the appropriate data. These programs are almost all based on the technique of free-energy minimisation which is derived from the following considerations.

**(i)** Problems related to the chemical equilibrium of a system involving several chemical species generally require the solution of many simultaneous nonlinear equations derived from a mass balance over all species coupled to the various equilibria between the species.

**(ii)** There is an infinite number of ways in which nonnegative mole numbers can be assigned to the possible products at equilibrium such that the chemical reactions involving the specified reactants will be balanced.

**(iii)** At a specified temperature and pressure, the most stable products (and their concentrations) are those associated with the lowest free energy of the system. That is, the problem can be reduced to that of minimising the free energy of the system.

The free energy of a multiphase, multicomponent system is given by,

$$G = \sum_i \sum_P n_i^p \mu_i^p$$

$$(3.112)$$

where $n_i^p$ = no of moles of species i in phase p.

$\mu_i^p$ = chemical potential of species I in phase p.

The chemical potential of species i is given by,

$$\mu_i^p = \mu_i^o + RT \ln a_i^p$$
$$= \Delta G_{f,i}^o + RT \ln a_i^p \qquad (3.113)$$

where $a_i^p$ = activity of species i in phase p and.

$\Delta G_{f,i}^o$ = Free energy of formation of species i at the specified temperature.

Each element is conserved in the reaction, so that,

$$\sum_i \sum_p \alpha_{ij} n_{ip} = \beta_j (j = 1, 2, \ldots, l)$$

(3.114)

where $\alpha_{ij}$ = no of atoms of element j in species i.

$b_j$ = no of moles of element j at start.

$l$ = total no of elements.

The problem becomes one of minimising Eq. (3.112) (substituted with Eq. (3.113) under nonstandard conditions) with the number of moles of each species in each phase as the variables, subject to the constraint of Eq. (3.114) as well as that of nonnegativity of the number of moles of each species present. The actual technique used to arrive at the minimum varies from program to program.

Most programs assume unit activity coefficients unless specified separately and several have the facility for making corrections for different temperatures using the Criss-Cobble Correspondence Principle.

## 3.7 Flow sheeting software

There are a number of software packages available that will carry out mass and energy balances around unit operations and more complex process flowsheets. Some of these are Metsim, Syscad, Aspen and, more recently, HSC for Chemistry. The user should be careful in using some of these in that most do not deal directly with ionic species but only neutral compounds. In the case of very simple chemistry, this does not create any problems but, in cases in which the metal ions are complexed, the results should be treated with caution.

Thus, the simple reaction for the oxidative leaching of the copper mineral covellite in chloride solutions by oxidation with copper(II) can be written as

$$CuS + CuCl_2(a) = 2CuCl(a) + S$$

(3.115)

This is the way that it would normally be written by engineers using one of these packages.

However, it really should be written as

$$CuS + Cu(II)(a) = 2Cu(I)(a) + S$$

(3.116)

because in the presence of chloride ions, both copper(II) and, particularly, copper(I) form several complexes with chloride ions and Eq. (3.115) does not specify which (or all) are being specified in the calculations.

Thus, if the thermodynamic data for $CuCl_2(a)$ and $CuCl(a)$ in Eq. (3.115) are for the undissociated aqueous species, then this equation will provide the thermodynamics of this reaction assuming that the only aqueous copper species are $CuCl_2(a)$ and $CuCl(a)$.

However, it is well known that in dilute chloride solutions up to about 1M, copper(II) is weakly complexed and $Cu^{2+}(a)$ is the predominant species while $CuCl_2^-(a)$ is the predominant copper(I) species.

Thus, to be more accurate, the above equation should be written as

$$CuS + Cu^{2+}(a) + 4Cl^-(a) = 2CuCl_2{\text{-}}(a) + S \qquad (3.117)$$

and this will reflect differently (than Eq. 3.115) on the thermodynamics of the process.

Thus, the standard Gibbs free energy change for reaction Eq. (3.115) can be calculated, to be $-44.03$ kJ mol$^{-1}$ while that for Eq. (3.117) is $-25.32$ kJ mol$^{-1}$ which is very different.

An accurate assessment of the thermodynamics of this reaction requires that all known complexes of copper(II) and copper(I) are included. Of course, at equilibrium, $\Delta G$ must have the same value irrespective of how we write the reaction. Thus, it should be the same for both Eqs (3.115) and (3.116) or any others. However, if we have a total copper(I) activity (or concentration) of say 0.1M, we cannot use this in the equilibrium calculations as the activity of $CuCl(a)$ or $CuCl_2^-(a)$ or $CuCl_3^{2-}(a)$ but will need the individual activities of each species. The same applies to all species taking part.

In many of these packages, $CuCl_2(a)$ is not the actual undissociated aqueous species but is simply used to denote a virtual species made up of $Cu^{2+}(a)$ and $Cl^-(a)$. Thus, the enthalpy (H) is simply the sum of that tabulated for $Cu^{2+}(a)$ and twice that for $Cl^-(a)$. Thus, these packages do not account for complexation of the metal ion unless a species such as, for example, NaCuCl$_2$(a) is listed which would take into account formation of the ion $CuCl_2^-(a)$. There are generally very few of such species available in the databases of these packages.

## 3.8 Summary

This Chapter has summarised the important concepts associated with the thermodynamics involved in the quantitative description of the reactions that occur in hydrometallurgical processes. The reader should be able to search for and manipulate appropriate thermodynamic data to enable an assessment of the thermodynamic viability of a particular reaction involving simple acid-base or complexation reactions including reaction with solid materials. The effect of temperature on the equilibria involved can, given available data, also be established.

The thermodynamics of reactions involving the transfer of electrons, that is, redox reactions have been dealt with in terms of electrode potentials. The reader should be able to manipulate potential data and use the Nernst equation to estimate potentials under nonstandard conditions. The effect of complexation of the metal ions on the electrode potentials is an important extension to the methodology that should now be available to the reader.

This Chapter has culminated in the introduction of oxidation state diagrams and, most importantly, the construction and interpretation of Eh/pH diagrams. The reader should be able to construct simple diagrams given suitable thermodynamic data and use the diagrams to predict whether certain reactions are possible and the conditions under which they are possible. The optimisation of the thermodynamics of reactions from a process point of view should now be possible for the reader.

## References

Bard, A.J., Parsons, R., Jordan, J. (Eds.), 1983. Standard Potentials in Aqueous Solution. Marcel Dekker, New York.

Bratsch, S.G., 1989. Standard electrode potentials and temperature coefficients in water at 298.15 K. J. Phys. Chem. Ref. Data 18.

Milazzo, G., Caroli, S., Braun, R.D., 1978. Tables of standard electrode potentials. J. Electrochem. Soc. 125, 261C.

Pourbaix, M., 1966. Atlas of Electrochemical Equilibria in Aqueous Solutions, English edition. Pergamon press, Oxford, p. 644.

## Further reading

Barin, I., Knacke, O., 1973. Thermochemical Properties of Inorganic Substances. Springer - Verlag, Berlin.

Barin, I., Knacke, O., Kubaschewski, O., 1977. Thermochemical Properties of Inorganic Substances (Supplement). Springer - Verlag, Berlin.

Barner, H.E., Scheuerman, R.V., 1978. Handbook of Thermochemical Data for Compounds and Aqueous Species. Wiley, New York.

Criss, C.M., Cobble, J.W., 1964a. The thermodynamic properties of high temperature aqueous solutions. V. The calculations of ionic heat capacities up to 200° and the correspondence principle. J.Amer. Chem. Soc. 86, 5385–5390.

Criss, C.M., Cobble, J.W., 1964b. The thermodynamic properties of high temperature aqueous solutions. V. The calculations of ionic heat capacities up to 200°. Entropies and heat capacities above 200°. J. Amer. Chem. Soc. 86, 5390–5393.

Jansz, J.J.C., 1983. Estimation of ionic activity in chloride systems at ambient and elevated temperatures. Hydrometallurgy 11, 13–31.

Nancollas, G.H., 1966. Interactions in Electrolyte Solutions. Elsevier, Amsterdam.

# Practice problems

These problems have been selected to provide the student of hydrometallurgy some practice in dealing with the calculations that are often required in a practical situation. One of the important tasks is to locate the appropriate thermodynamic data required to solve the problem. Many of these problems require the student to locate such data primarily in the Appendices to Chapter 2.

## Hints in solving thermodynamic problems

In most problems involving chemical thermodynamics, you can follow these rather general steps.
- Write down the balanced chemical reaction(s) involved.
- Write out the equilibrium quotient(constant) expression.
- If required, calculate the equilibrium constant from the thermodynamic data provided or in the tables. This generally involves calculating the $\Delta G$ of the reaction and then calculating K from it.
- Try and solve for the unknown quantity in the equilibrium expression by substituting known(or given) activities. This may involve simple mass balances of the important species.
- Valid approximations can often simplify the calculations that can often be carried out by simple iterative methods.
1. Calculate the solubility of AgCl (in mol L$^{-1}$) in a solution containing 2 mol L$^{-1}$ NaCl. Assume unit activity coefficients.

$$\mathrm{AgCl(s)} = \mathrm{Ag^+ + Cl^-}; \mathrm{p}K_{sp} = 9.75$$

$$\mathrm{Ag^+ + 2Cl^-} = \mathrm{AgCl_2^-}; \mathrm{Log}\ \beta_2 = 5.25$$

2. Alumina ($Al_2O_3$) dissolves in concentrated alkaline solutions according to the equation:

$$0.5Al_2O_3(s) + OH^- + 1.5H_2O = Al(OH)_4^-$$

with an equilibrium constant of 4 at 25°C. Calculate the solubility of alumina at pH = 10. Express your answer in mol dm$^{-3}$ and kg m$^{-3}$.

3. Given the following equilibrium constants at 25°C

$$CO_3^{2-} + H_2O = HCO_3^- + OH^- \quad K = 10^{-3.7}$$

$$H_2O = H^+ + OH^- \quad K_w = 10^{-13.99}$$

$$Al(OH)_3(s) = Al^{3+} + 3OH^- \quad K_{Sp} = 10^{-32.34}$$

$$Al^{3+} + 4OH^- = Al(OH)_4^- \quad \beta_4 = 10^{32.99}$$

(a) Calculate the pH of a solution which contains the two salts $Na_2CO_3$ (250 g $L^{-1}$) and $NaHCO_3$ (2 g $L^{-1}$) at equilibrium at 25°C.

(b) Calculate the solubility of gibbsite $Al(OH)_3(s)$ in a solution of pH 12.5 at 25°C and express your answer in kg $m^{-3}$.

(c) Clearly mention any assumptions that you made in the calculations and suggest what factors or equations you would consider improving the reliability of your answer.

4. Based on the solubility product of uranyl hydroxide $UO_2(OH)_2$ and the equilibrium constant for the formation of uranyl sulphate complex $UO_2(SO_4)$ (aq) in Table 2.4 of Chapter 2, predict the concentration of uranium dissolved as $UO_2(SO_4)$ in equilibrium with $UO_2(OH)_2$ in a solution of 0.1 M sulphate maintained at pH 7. Repeat your calculation assuming that the dissolved uranium is in the form $UO_2(SO_4)_2^{2-}$.

5. A solution contains the two metal chlorides $FeCl_2$ and $MnCl_2$ at pH 3. The concentration of each salt is 0.1 mol $L^{-1}$ and the solubility products of the two hydroxides are as follows:

$$Fe(OH)_2(s) = Fe^{2+} + 2OH^- \quad pK_{sp} = 15.1$$

$$Mn(OH)_2(s) = Mn^{2+} + 2OH^- \quad pK_{sp} = 12.8$$

(i) Predict the pH at which each metal ion will precipitate as the hydroxide $M(OH)_2$ at 298 K.

(ii) Predict which metal ion would precipitate first (as the hydroxide) upon the addition of NaOH to the above solution.

(iii) What percentage of the first metal ion will precipitate before the commencement of the precipitation of the second metal ion?

6. Ammonium uranyl carbonate (AUC) with the molecular formula $(NH_4)_4UO_2(CO_3)_3$ is an important intermediate material in the production of oxides for the nuclear fuel cycle. The

standard free energy change ($\Delta G^\circ$) for the precipitation of AUC according to the reaction:

$$UO_2(NO_3)_2.6H_2O(Aq) + 6NH_3(g) + 3CO_2(g)$$

$$\rightleftharpoons (NH_4)_4UO_2(CO_3)_3(s) + 2NH_4NO_3(aq) + 3H_2O(l)$$

is $-20$ kJ at 60°C. Calculate the percentage of uranium that would precipitate from a solution containing 300 g L$^{-1}$ of uranium if the equilibrium was attained by sparging pure ammonia and carbon dioxide gases at a volume ratio of $NH_3/CO_2 = 2/1$, atmospheric pressure, and 60°C. State any assumptions made in your calculation.

7. Fill in the blanks (shown by question marks) in the following table and predict with reasons whether the $Au^+$ ion is stable in aqueous media at 25°C.

| Electrode reaction | $E^\circ$/V | $\Delta G^\circ$/kJ mol$^{-1}$ |
|---|---|---|
| $Au^+ + e = Au(s)$ | 1.69 | ? |
| $Au^{3+} + 3e = Au(s)$ | 1.50 | ? |
| $Au^{3+} + 2Au(s) = 3Au^+$ | | ? |

8. Calculate the **change in Eh** of the following reaction, when the pH is changed from 5 to 8 at 298 K.

$$2Cu(OH)_2 + 2H^+ + 2e = Cu_2O + 3H_2O \quad E^\circ = -0.065 \text{ V}$$

9. A sulphate leach solution contains 50 g L$^{-1}$ of copper and 30 g L$^{-1}$ of ferric ions. Use the hydroxide precipitation diagram to estimate

   (i)   the pH at which copper and iron start to precipitate as their hydroxides at 25°C.

   (ii)  the concentration of iron in solution at the pH when 1% of the copper is precipitated.

   (iii) the pH when 99.9% of the copper has precipitated.

10. Most hydrometallurgical processes for base metals are carried out in sulphate solutions. Lead is often an impurity which forms insoluble $PbSO_4$ which reports to the leach and other residues. This poses an environmental problem in terms of safe disposal of such residues.

   You are given the problem of removing lead from such a residue and it is suggested that lead sulphate can be dissolved in acetate solutions because the lead ion forms a relatively stable acetate complex.

You locate the following data on the solubility of lead sulphate in water and the stability constant for the formation of the di-acetate complex,

$$K_{sp} \text{ (PbSO}_4) = 1. \, 10^{-8}$$

$$\beta_2 = 10^6 \text{ for Pb}^{2+} + 2Ac^- = Pb(Ac)_2 \text{ (aq)}$$

(a) Calculate the solubility of lead in a solution containing 1.0 mol $L^{-1}$ of acetate ions.

(b) Given that $pK_a = 4$ for acetic acid, what would be the effect of pH on the solubility and recommend a suitable operating pH for the dissolution process.

(c) Suggest a possible method for recovering lead from the solution.

11. **(Difficult)** A typical electrolyte in the electrowinning of zinc contains 1.2 mol $L^{-1}$ zinc at a pH of 1.5 at 25C.

The following ion-pair equilibria exist in this solution

$$Zn^{2+}(aq) + SO_4^{2-}(aq) = ZnSO_4(aq) \quad K = 300 \text{ L mol}^{-1} \text{ at 25C}$$

$$H^+(aq) + SO_4^{2-}(aq) = HSO_4^-(aq) \quad K = 100 \text{ l mol}^{-1} \text{ at 25C}$$

(a) Determine the ionic strength of the solution assuming all activity coefficients to be unity. You may assume that uncharged $H_2SO_4(aq)$ is present in negligible amounts.

The following extended form of the Debye-Huckel equation can be used to describe the activity coefficient($\gamma$) of an ionic species,

$$\log_{10}\gamma = 0.5 \, z^2 \, \sqrt{I} \, / (1 + 0.33\phi\sqrt{I})$$

where z is the ionic charge, I the ionic strength (with concentrations in mol $L^{-1}$) and the parameter $\phi$ is 9 for $H^+$, 6 for $Zn^{2+}$ and 4 for $SO_4^{2-}$ and $HSO_4^-$.

Activity coefficients for neutral aqueous species may be assumed as unity.

(b) Calculate activity coefficients for the ions present and recalculate I. This iterative procedure can be repeated to estimate improved values for the activity coefficients.

(c) Calculate the reduction potential for the reaction

$$Zn^{2+}(aq) + 2e = Zn(s)$$

at 25C in the above electrolyte.

12. In the electrowinning of zinc from sulphate solutions, small amounts of magnesium accumulate in the electrolyte and must be periodically removed. Using the data in the

Appendices, assess the following method which will permit the separation of the magnesium from the zinc by selective precipitation. The precipitation will be carried out by adding lime(Ca(OH)$_2$) to the spent electrolyte which has a composition of 65.4 g L$^{-1}$ zinc, 24.3 g L$^{-1}$ magnesium and 98 g L$^{-1}$ sulphuric acid. You should demonstrate that this is possible by calculating the solubility of each metal ion at pH values of 6.0 and 7.0.

Estimate the rate of addition of lime (tonnes hour$^{-1}$) required to control the pH at a value of 7.0 for a stream of electrolyte having the above composition at a flowrate of 1 m$^3$ hour$^{-1}$ at 25°C. Assume unit activity coefficients for all species.

**13.** Calculate the lowest pH at which the activity of nickel ions can be lowered to 0.01 molar by reducing at 298K a solution of nickel sulphate, with hydrogen gas maintained at a partial pressure of 20 atmospheres assuming ideal gas and solution behaviour.

$$Ni^{2+} + 2e = Ni \quad E^o = -0.241V.$$

What would happen to the pH of the solution during the reduction process and suggest how would you control it to the above calculated value?

**14.** Nickel metal is produced in by the reduction of the tetraamine complex Ni(NH$_3$)$_4^{2+}$ by hydrogen gas under pressure. Calculate the equilibrium hydrogen pressure(in atmospheres) required to precipitate 90% of the nickel from a solution containing 1.0 mol L$^{-1}$ nickel as the above complex and 1.0 mol L$^{-1}$ free ammonia(NH$_3$) at a pH of 11 at 25°C. Assume ideal solutions and gases.

$$Ni^{2+} + 2e = Ni \quad E^o = -0.24 \text{ V}$$

$$Ni^{2+} + 4 NH_3 = Ni(NH_3)_4^{2+} \quad \log \beta_4 = 7.7$$

Suggest two changes to the operating conditions which would increase the degree of precipitation.

**15.** A solution from the ammonia/ammonium sulphate leaching of a nickel sulphide ore contains 50 g L$^{-1}$ Ni$^{2+}$ and 2 g L$^{-1}$ Co$^{2+}$. Nickel is to be precipitated from solution as a metal powder by hydrogen under pressure. Calculate the theoretical maximum recovery of nickel containing less than 0.01 mole per cent cobalt that can be obtained from this solution. $\Delta G^o$ (Ni$^{2+}$) = $-46.6$ kJ mol$^{-1}$ and $\Delta G^o$ (Co$^{2+}$) = $-51.5$ kJ mol$^{-1}$.

**16.** Calculate the maximum concentration of copper(in mol dm$^{-3}$) which can be achieved by dissolving an ore consisting of CuO in aqueous sulphuric acid solutions at a controlled pH of 3.5 at 70°C using the thermodynamic data from the Appendices and the following for aqueous $CuSO_4$ solutions

| Ionic strength | Mean activity |
|---|---|
| Mol L$^{-1}$ | Coeff. |
| 0.1 | 0.150 |
| 0.5 | 0.067 |
| 1.0 | 0.046 |
| 2.0 | 0.037 |

What would be the corresponding solubility at a pH value of 4.5?

17. Gold forms a very stable complex with the cyanide ion, $Au(CN)_2^-$, which has a stability constant of $10^{40}$. A gold leaching process uses a cyanide ion concentration of $1.10^{-3}$ mol L$^{-1}$ at a pH of 10.5.

    (a) Calculate the concentration of the uncomplexed gold ion ($Au^+$) in a leach solution containing 2 ppm(mg L$^{-1}$) of dissolved gold.

    (b) The stability constant for the protonation of the cyanide ion is $10^9$. Calculate the concentration of the unprotonated cyanide ion at a pH of 10.5 in a solution containing $1.10^{-3}$ mol L$^{-1}$ of total cyanide.

    (c) Sketch the expected relationship between the fraction of cyanide in the unprotonated form as a function of the pH.

    (d) Calculate the equilibrium potential of a gold electrode in the above leach solution given that $E_0 = 1.68V$ for the reaction

$$Au^+ + e = Au$$

18. Copper(I) forms stable cyanide complexes which can consume cyanide in the cyanidation process for the recovery of gold. The most important are $CuCN$ and $Cu(CN)_2^-$ for which the following data at 25°C applies,

$$CuCN(s) = Cu^+(aq) + CN^-(aq) \ K_{sp} = 10^{-10}$$

$$Cu^+(aq) + 2CN^-(aq) = Cu(CN)_2^-(aq) \ \beta_2 = 10^{12.5}$$

Calculate the concentration of soluble copper(I) ions in a solution containing $0.001$ mol $L^{-1}$ free $CN^-$ which is saturated with CuCN at 25°C.

Calculate the solubility of copper(I) in a solution containing $0.001$ mol $L^{-1}$ total cyanide(as $CN^- + HCN$) at a pH value of 8.3 if the equilibrium constant for the protonation of $CN^-$

$$CN^-(aq) + H_3O^+(aq) = HCN(aq) + H_2O$$

is $10^{9.3}$ at 25°C.

19. An ammoniacal sulphate solution resulting from the ammonia leaching of a nickel sulphide concentrate contains $58.7$ g $L^{-1}$ $Ni^{2+}$ and $5.89$ g $L^{-1}$ $Co^{2+}$. Nickel is recovered by reduction using hydrogen gas at 100 atmospheres(bar) at 25°C. Assuming that nickel and cobalt are present as the aquo-ions,
   **(a)** Calculate the values for the equilibrium constants for each reaction at 25°C.
   **(b)** Calculate the maximum recovery of nickel free of cobalt that can achieved from this solution.
   **(c)** Calculate the minimum pH which can be used to achieve this recovery.
   **(d)** If the stability constant for the $Ni(NH_3)_4^{2+}$ ion is $10^8$ and that for the $Co(NH_3)_4^{2+}$ is $10^7$, describe qualitatively the effect of such complexation on the results of the calculations (ii) and (iii) above.
   Assume unit activities for all dissolved species.

20. A zinc sulphate solution contaminated with 0.1M each of $Fe^{2+}$, $Cu^{2+}$ and $Cd^{2+}$ is purified by oxidation of $Fe^{2+}$ to $Fe^{3+}$ by oxygen, recovery of $Fe(OH)_3$ by precipitation with NaOH and cementation (reduction) of the remaining $Cu^{2+}$ and $Cd^{2+}$ from solution with zinc metal dust.
   **(iv)** Calculate the ratio of $Fe^{2+}/Fe^{3+}$ after oxidation with oxygen at 0.2 atmosphere(i.e. air) at pH 1 and at pH 3.
   **(v)** Calculate the pH at which 99% of the iron is precipitated as $Fe(OH)_3$. What % of the copper is precipitated at this pH?
   **(vi)** Calculate the concentration ratio of $Cu^{2+}/Zn^{2+}$ and $Cd^{2+}/Zn^{2+}$ remaining in solution after cementation with excess zinc dust.

21. A version of the $E_H$-pH diagram for the Cu-O system at 25°C with $10^{-3}$ mol $L^{-1}$ cupric ions is given in Fig. 3.16.
   **(a)** Write balanced equations and derive the equations which define the relationship between $E_H$ and pH for all lines.

**(b)** It is suggested that copper metal could be produced by reduction of cupric ions by hydrogen. Is this feasible? Why?

**(c)** What is the stable product of the corrosion of copper at pH 8? Why?

**(d)** Show that it is thermodynamically possible to leach copper metal in aerated acid solutions.

**(e)** Which thermodynamic property of a metal would one require to assess whether the metal could cement copper from an acidic solution and what value would it have to have?

22. The $E_H$/pH diagram for the iron/oxygen/ammonia system at 25°C is given in the Appendix. Answer the following questions by using the information in the diagram.

**(a)** Will metallic iron spontaneously dissolve in a **deoxygenated** ammonia/ammonium ion solution at pH 5, pH 9.5, pH 12? If so, write balanced equations for the reactions and justify your answers.

**(b)** Metallic iron is placed in an **aerated** ammonia/ammonium ion solution at pH 9.5. What is the stable product of reaction and how would you expect the rate of reaction to vary with time? Justify your answers.

**(c)** Estimate the stability constant for the formation of the iron(II) diamine complex ion.

23. Using the partial $E_H$-pH diagram in the Appendix for the Fe-S-$H_2O$ system at unit activity soluble species at 298K, answer the following:

**(a)** Shade in the area of stability of elemental sulphur.

**(b)** What are the products of oxidation of pyrite at pH 1, pH 8 and pH 14?

**(c)** Write a balanced equation for the leaching of pyrite using ferric ions at a pH value of 1.

**(d)** Estimate the $\Delta G°$ for the reaction at 298K.

**(e)** How would the region of stability of the ferric ion change with increasing temperature?

**(f)** Suggest an alternative method for the leaching of pyrite and give a balanced chemical reaction for it.

**(g)** A sample of pyrrhotite is exposed to a moist environment at a pH of about 6 for several months. What would you expect to see if the sample were sectioned and examined microscopically?

24 With reference to the $E_H$/pH diagrams of the Ni-S-water system in the Appendix and the S-water system at 25°C, answer the following questions.

**(a)** Write out balanced chemical equations for lines 1, 2 and 3

**(b)** Estimate the equilibrium constant for the reaction

$$Ni^{2+} + 2H_2O = Ni(OH)_2 + 2H^+$$

**(c)** Nickel(III) oxide can be produced by anodic oxidation of nickel(II). Suggest with reasons possible conditions under which this reaction can be carried out. What competing reaction could be expected at the anode?

**(d)** You are asked to develop a process which will leach a nickel sulphide concentrate containing mainly heazlewoodite ($Ni_3S_2$). Suggest with reasons and balanced equations how this could be accomplished and use the diagram to justify your process from a thermodynamic point of view. What would be the fate of the sulphur in the concentrate after your leach step? A cross section of an incompletely leached grain of the concentrate is examined microscopically. What would you expect to see?

**(e)** Nickel metal can be recovered from solution by electrowinning. Use the diagram to show that this is thermodynamically not the favoured reaction at a cathode in acidic solutions and give reasons why it can nevertheless be achieved. What would you do to maximise the current efficiency for the deposition of the metal? What would be the minimum cell voltage for the electrowinning process at pH 3?

**25.** The $E_H$-pH diagram, summarises the thermodynamics of the U-$H_2O$-$CO_2$ system for a total carbonate concentration of 4 mol $L^{-1}$.

**(a)** What is the oxidation state of U in $UO_2$, $UO_2^{2+}$, $UO_2(CO_3)_3^{4-}$ and $UO_2(OH)_2$ ?

**(b)** Indicate by hatching the relevant area, the regions of stability of the solid phases at a total uranium concentration of 2.38 g $L^{-1}$.

**(c)** Suggest, with reasons, a method and appropriate operating conditions for leaching an ore containing $UO_2$ to produce the tri-carbonate complex.

**(d)** Can $UO_2$ be precipitated from this leach solution by reduction with hydrogen under pressure? If yes, suggest, with reasons, optimum thermodynamic conditions for this reaction.

**26.** The potential of a silver electrode which consists of a silver wire in equilibrium with 0.05 mol $L^{-1}$ $AgNO_3$ is 0.517 V with respect to a reference electrode at 298 K. The potential of the reference electrode with respect to the SHE is 0.200 V

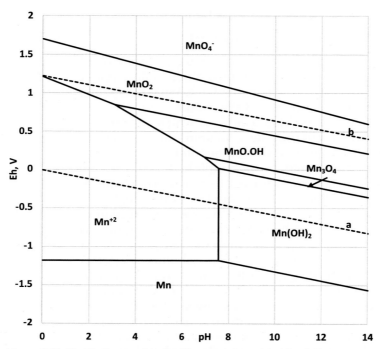

**Figure 3.17** Eh-pH diagram for the Mn-water system 0.1M soluble species.

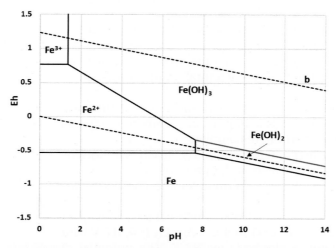

**Figure 3.18** Eh-pH diagram for the Fe-water system 0.01M soluble species.

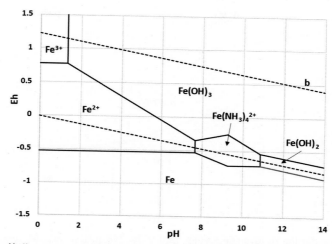

**Figure 3.19** Eh-pH diagram for the Fe-ammonia-water system. 0.01M Fe species, 3 M $NH_3 + NH_4^+$.

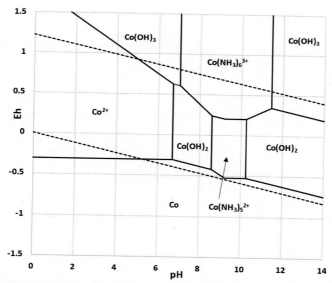

**Figure 3.20** Eh-pH diagram for the Co-ammonia-water system. 0.1M Co species, 2 M $NH_3 + NH_4^+$.

and $E^o$ $(Ag^+/Ag) = 0.799$ V at 298 K. Calculate the activity and activity coefficient of $Ag^+$.

27. In the roast-leach-electrowin process for the production of zinc metal, zinc metal dust is added to the acidic liquor to remove impurities such as cadmium, nickel and cobalt. Write

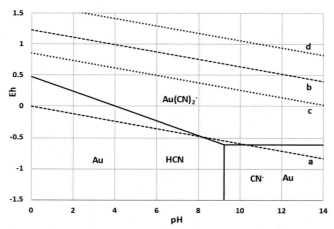

**Figure 3.21** Eh-pH diagram for the Au-Cyanide system. $[Au] = 10^{-4}$ M, $[CN] = 10^{-2}$ M. Line "c" is for $O_2/H_2O_2$ equilibrium and line "d" for $H_2O_2/H_2O$ equilibrium.

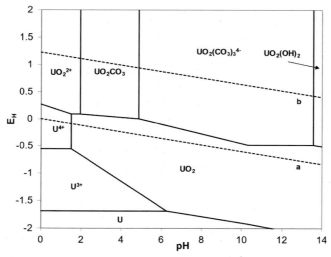

**Figure 3.22** Eh-pH diagram for the U-carbonate-water system. $10^{-2}$M soluble U and 1M $CO_3^{2-} + HCO_3^-$.

the relevant equation for cobalt and calculate the concentration of $Co^{2+}$ in equilibrium with 0.8 mol $l^{-1}$ $Zn^{2+}$ at 298 K.

28. Fill in the blanks (shown by question marks) in the following table and predict with reasons whether the $Cu^+$ ion is stable in aqueous sulphate media at 25°C.

Calculate the equilibrium constant for the reaction $Cu^{2+} + Cu = 2Cu^+$.

Calculate the E° for the redox couple: $CuCl_2^- + e = Cu(s) + 2Cl^-$, assuming

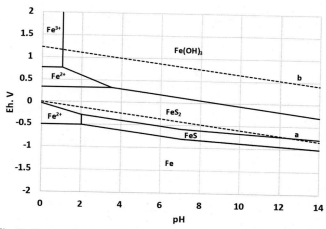

**Figure 3.23** Eh-pH diagram for the Fe-S-water system. 0.1M soluble Fe species, no sulphate.

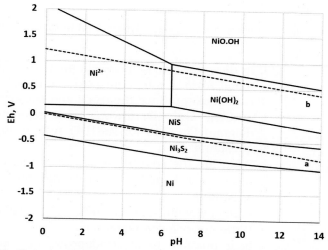

**Figure 3.24** Eh-pH diagram for the Ni-S-water system. 0.1M soluble Ni species, no sulphate.

Log $B_2 = 5.5$ for $Cu^+ + 2Cl^- = CuCl_2^-$.

| Electrode reaction | $E^o$/V | $\Delta G^o$/kJ mol$^{-1}$ |
|---|---|---|
| $Cu^+ + e = Cu(s)$ | 0.52 | ? |
| $Cu^{2+} + 2e = Cu(s)$ | 0.34 | ? |
| $Cu^{2+} + Cu(s) = 2Cu^+$ | | ? |

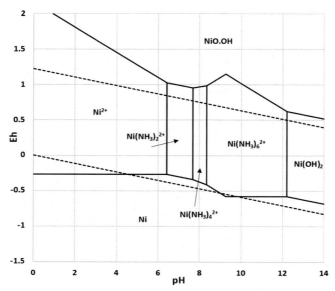

**Figure 3.25** Eh-pH diagram for the Ni-ammonia-water system. 0.1M Ni species, 2 M $NH_3 + NH_4^+$.

**29.** In the roast-leach-electrowin process for the recovery of zinc metal from a sphalerite (ZnS) concentrate, the sulphide is roasted to produce the oxide and subsequently leached and purified and electrowon. With reference to the Eh-pH diagram Fig. 3.10 for Zn-$H_2O$ system answer the following questions.

What are the conditions at which ZnO can be dissolved to produce a 1 $mol\,L^{-1}$ Zn(II) solution? Write the relevant equations.

Predict whether the zinc metal is stable in water at pH 0, 7, 13. Write out relevant equations.

Is it possible to electrodeposit zinc metal from aqueous solutions? Justify your answer.

**30.** Zinc metal is recovered using electrolysis of an acidic solution of $ZnSO_4$ solution using a zinc cathode and a lead (inert) anode.

Calculate the cell voltage at standard conditions using the following data. State the assumptions made in the calculation.

$$Zn^{2+}(aq) + H_2O(l) = Zn(s) + 0.5O_2(g) + 2H^+(aq) \text{ (overall cell reaction)}$$

$$0.5O_2(g) + 2H^+(aq) + 2e = 2H_2O; \; E^o = 1.23 \text{ V}$$

$$Zn^{2+}(aq) + 2e = Zn(s); \quad E^o = -0.76 \text{ V}$$

What would be the cell voltage if the activity of $Zn^{2+}$ dropped to 0.01 when the other conditions remained the same?

**31.** Calculate the values of $\Delta G^o$ and the equilibrium constant K at 298 K for the reaction between $Au^{3+}$ and Au using the data in the Table below. In your answer show the calculated values for the thermodynamic quantities indicated by question marks in the Table.

Is $Au^+$ stable in an aqueous solution? If your answer is "yes" give reasons. If your answer is "no" suggest with reasons what you would add to prepare a solution of Au(I).

**32.** Calculate the concentration of Cu(II) in the following systems using the thermodynamic data in the Table below.

**(i)** In water, saturated with $Cu(OH)_2$ at pH 10 and 298 K ($pK_w = 14$).

**(ii)** In a concentrated ammonia solution saturated with $Cu(OH)_2$ at pH 10 and 298 K if the $NH_3$ concentration remains at 2 M. State the assumptions made in the calculation.

| Reaction | E°/V | $\Delta G^o$/J mol$^{-1}$ at 298 K | K |
|---|---|---|---|
| *Data for Q 49* | | | |
| $Au^+(aq) + e = Au(s)$ | 1.69 | $\Delta G^o$ ? | |
| $Au^{3+}(aq) + 3e = Au(s)$ | 1.50 | $\Delta G^o$ ? | |
| $Au^{3+}(aq) + 2Au(s) = 3Au^+(aq)$ | | $\Delta G^o$ ? | K ? |
| *Data for Q 50* | | | |
| $Cu(OH)_2(s) = Cu^{2+}(aq) + 2OH^-(aq)$ | | | $10^{-20}$ |
| $Cu^{2+}(aq) + 3NH_3(aq) =$ $Cu(NH_3)_3^{2+}(aq)$ | | | $10^{11}$ |

# Case study 1

You are a metallurgist at an electrolytic zinc plant which has the following overall simplified flowsheet.

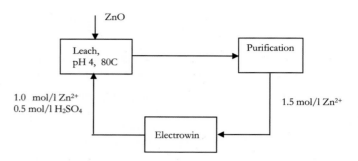

You are asked to assess whether it is possible to conduct the leach at a pH of 5 instead of 4 as there could be advantages in operating the purification process for the electrolyte at the higher pH. Using the data below and in the Appendices, compile a brief report on the technical feasibility of the proposed change in the pH and what the implications of the change will make on the concentration of zinc in the leached solution. What effect could the change in operating pH have on the rate of leaching? (Hint: Make your initial calculations assuming all species have unit activity coefficients).

| Ionic strength | Mean ionic activity coefficient $H_2SO_4(aq)$ | Mean ionic activity coefficient $ZnSO_4$ (aq) |
|---|---|---|
| 1.0 | 0.131 | 0.044 |
| 2.0 | 0.125 | 0.035 |
| 4.0 | 0.172 | 0.037 |
| 6.0 | 0.190 | 0.053 |

## Case study 2

Sea water contains approximately 0.05 mol L$^{-1}$ Mg$^{2+}$ and 0.01 mol L$^{-1}$ Ca$^{2+}$. Magnesia can be recovered from sea water by reaction with calcined seashells by the following reaction

$$Mg^{2+} + CaO + H_2O = Mg(OH)_2 + Ca^{2+}$$

Using thermodynamic data in the Appendices, calculate the concentration of magnesium ions at equilibrium in a batch of sea water in contact with excess lime at 298K.

**(a)** Calculate K for the reaction at 25°C.

**(b)** Calculate a species distribution diagram for the addition of increasing amounts of CaO to a fixed volume of sea water.

**(c)** Now include Ca(OH)$_2$ in the calculations. What differences do you notice?

(d) What could be the effect of including the monochloro-complexes of $Mg^{2+}$ and $Ca^{2+}$ in the calculations assuming that sea water contains 0.5M NaCl.

## Case study 3

(a) Construct an $E_H$ - pH diagram (pH 0–14) for the Ni-$H_2O$ system at 25°C and $10^{-3}$ M activities for all soluble species given the following thermodynamic data,

$$Ni^{2+} + 2e = Ni \quad E^o = -0.26V$$

$$NiOOH + 3H^+ + 3e = Ni + 2H_2O \quad E^o = 0.52V$$

$$Ni(OH)_2(s) = Ni^{2+} + 2OH^- \quad K_{sp} = 6. \, 10^{-16}$$

Areas of stability for all of the above Ni species should be shown.

Also include lines of stability for water.

(b) NiOOH is the active cathode(on discharge) component in NiCad batteries which operate with the anodic couple

$$Cd(OH)_2 + 2H^+ + 2e = Cd + 2H_2O \quad E^o = 0.003V$$

Draw this line on the diagram and predict the open circuit voltage of a NiCad and the effect of the pH on it.

(c) A well-known process for removing soluble Co from Ni sulphate solutions is to oxidise Co with NiOOH. Assess the thermodynamic feasibility of this process given

$$Co(OH)_3 + e = Co(OH)_2 + OH^- \quad E^o = 0.17V$$

$$Co(OH)_2 = Co^{2+} + 2OH^- \quad K_s = 1. \, 10^{-15}$$

How would one make NiOOH and under what conditions would one carry out the Co removal step?

## Appendix

The following $E_H$/pH diagrams of some common systems will prove to be useful in both answering the problems and in interpreting the chemistry of leaching, separation and reduction processes which will be dealt with in later chapters (Figs. 3.17–3.25).

# 4

# Material and energy balances

It is a fundamental requirement that what goes into a metallurgical plant must come out. There may be a number of inputs and outputs into the overall flowsheet, but everything that enters into the process plant comes out in one stream or another unless it accumulates in the plant which is very unusual. One example of an accumulating material is fine gold particles in the liners of a mill which is only recovered when the mill is relined-a period which can be several years.

The determination of the deportment of the materials input into a plant is an integral part of plant design. For each unit operation, the size and concentration of each output stream needs to be known so that the optimum size of equipment, such as pumps, tanks, and other equipment, can be specified.

The ultimate aim of a mass balance in a plant is to know exactly where all of the material is located within the plant at any point in time. This information can then be used for several different purposes depending upon the stage of the project such as

- Plant design
  - where does the mass report?
  - sizes of equipment
    - holding tanks, reactors, cells, thickeners and so on
  - flow rates
    - pumps, pipes, flow gauges
- Control strategies
  - automatic
    - feed forward, feed back
    - PID
  - manual
    - rare, but still useful
- Determination of the overall efficiency of the plant, that is, the recovery of the valuable metal

Hydrometallurgy. https://doi.org/10.1016/B978-0-323-99322-7.00006-5

## 4.1 Material balances

### 4.1.1 Units and analyses

In considering a mass balance, it is important to be consistent with the units being used and to use one measure of mass throughout. The most common units used for concentration are those provided by analytical laboratories and, depending on the concentration, are expressed as mass %, oz tonne$^{-1}$, g tonne$^{-1}$ or parts per million, ppm (the last two are equivalent). For solutions, the most common unit is g L$^{-1}$ although ppm (mg L$^{-1}$) is also often used for low concentrations.

Gold is typically described and sold by the ounce (oz); however, these are not the Avoirdupois ounces but Troy ounces which are equivalent to 31.103 g. The concentrations of some metals are for historic reasons quoted in terms of the final product. Thus, uranium is usually considered as $U_3O_8$ so a solution containing 100 g U L$^{-1}$ is the same as a solution containing 117.9 g $U_3O_8$ L$^{-1}$.

An analysis of a phosphate ore used in the production of fertiliser is given in Table 4.1.

This particular analysis has several interesting things to note...

- The units are mixed according to level, high levels are mass %, low levels are in ppm (g t$^{-1}$).
- The accuracy of analysis of the main three elements (P, Ca, S) is given to three significant figures whilst the remainder are given to only two.
- Several elements (e.g. Bi, Hg) are below the detection limit of the analytical equipment used, the detection limit is different for each element.

**Table 4.1 Analysis of a phosphate ore.**

| Element | ppm | Element | ppm |
|---------|-------|---------|-------|
| P | 13.0% | Hg | <2 |
| Ca | 31.0% | Mg | 0.24% |
| S | 1.50% | Mo | 4.1 |
| Fe | 0.17% | Pb | 3 |
| Mn | 24 | Sb | <2 |
| As | 4 | Se | 10 |
| Bi | <20 | Sr | 0.18% |
| Cd | 26 | Tl | <2 |
| Cr | 140 | V | 36 |
| Cu | 7.7 | Zn | 16 |

- The total mass is less than 100% as there are no data for common elements such as C, O, N, H. The only nonmetals of interest are those of major commercial interest, that is P, those whose trace presence is desirable in fertiliser such as Cu and Mo and those of potential environmental concern such as As and Hg.

## 4.1.2 Mass and flow balances

There are two major types of balance, a mass balance and a mass flow balance, the principles of these are the same but the outcome may be somewhat different.

A mass balance is simply the balance of the masses in a process regardless of any time constraints, that is given enough time the mass input into a system will come out. This describes a **batch** operation where the process is operated until it is completed, and the resulting material discharged from the batch reactor as shown in Fig. 4.1.

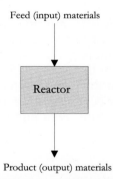

Feed (input) materials

Reactor

Product (output) materials

**Figure 4.1** Batch or continuous reactor.

However, batch processes tend to be labour-intensive with attention needed at the start and end of every batch. Sequences of batch operations often cause problems within plants as the time taken for each batch will be different and storage is required at every stage to ensure that the correct mixtures are made at the various subsequent batch operations. An advantage of such processes, however, is that accounting for the materials in the process is considerably easier and more accurate than in continuous processes, and this approach is generally used in precious metal refineries.

The simple example above can be converted it into a flowsheet containing a number of unit operations for carrying out the overall process. In some cases, one or more operations will be conducted simultaneously, and these are known as parallel operations and can be performed completely independently of each other. An example could be the management of anodes and cathodes in an electrowinning tankhouse.

The mass flow balance is slightly different in that it is the material flow per unit time which is used. At any time, the flow of material in may be more or less than the flow of material out, or the same. If the flow in is greater than that out, then accumulation occurs in the plant, whereas, if the flow in is smaller, then the plant becomes progressively emptier as there is a net outflow.

When the rate of input of materials equals the rate of output, the plant processes material at a constant rate and the system is in a steady state. The optimum way to operate a plant is under steady-state conditions as this requires the least control. A plant where the flows continuously fluctuate requires constant adjustments which can require storage facilities if needed to ensure each operation has its optimum flow and works at its greatest efficiency.

Where possible, processes are operated continuously under steady-state conditions as this requires the least attention and control is aimed at retaining the status quo. Batch processes are generally used where there is insufficient flow of material to operate continuously, or the operation has a series of steps which cannot be practically or economically separated.

Examples of batch and continuous processes in hydrometallurgy are

| Batch | Continuous |
|---|---|
| Ion exchange | Gold/tank leaching |
| Heap leaching | Solvent extraction |
| Stripping gold from carbon | Electrowinning |

## 4.1.3 Residence time

In all systems, each vessel has a fixed volume and is therefore restricted in the amount it can process at any given time. For a batch system, the limitation is obvious—you cannot put in more than it can hold. A continuous flow reactor can be thought of as like a garden hose, water flows in one end from the tap and out the other end at the same rate (i.e., steady-state) the time spent travelling from the tap to the garden is a function of the diameter of the hose and the water flowrate. The water in the the hose is replaced periodically. Thus, a 30 m hose of 20 mm diameter contains $(3000 \times 2^2 \times \pi/4/1000) = 9.425$ L of water, and at a flowrate of 10 L min$^{-1}$ the water is changed every 56.6 s. The time taken to completely change the volume of a vessel is called the *mean residence time*.

In general, *mean residence time* = volume/flowrate.

Some processes require a certain residence time for completion, for example a copper heap leach may have a residence time of several months, but a solvent extraction mixer may only need 30 s.

When designing a plant, it is necessary to know the effect of various residence times on the process so that the optimum size of vessel can be specified for the flowrate. Gold recovery may be only partially complete after 24 h of leaching but the extra gold gained by increasing the residence time to 48 h may not be sufficient to cover the cost of constructing, maintaining, and operating a significantly larger plant.

## 4.1.4 Mineral processing plants

Mineral processing operations provide relatively simple examples of mass balances as there are no phase changes, no chemical reactions and generally no changes in temperature. For a complete plant, the simplest form of mass balance is shown in Fig. 4.2.

Obviously, the material fed into the plant must equal the material leaving the plant,

$$m_{feed} = m_{conc} + m_{tail}$$

in which m refers to mass. Similarly, the mass flow balance can be written

$$L_{feed} = L_{conc} + L_{tail}$$

in which L refers to the mass flowrate.

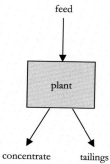

**Figure 4.2** Process streams in overall mineral processing plant.

However, since the mass flow is time dependent, there may come a time where the feed or one of the streams out of the process is interrupted and a change in the net flow occurs and material either accumulates or diminishes in the plant.

$$\text{Net flow} = L_{\text{feed}} - (L_{\text{conc}} + L_{\text{tail}})$$

These equations deal with the total mass within the system, and this can be broken down into a number of separate contributions which must also be balanced. In mineral processing plants there is no conversion from one component to another and the mineral phases entering the plant also leave the plant unchanged.

The major items which need to be balanced are
◆ water
◆ individual component minerals such as quartz, sphalerite, gold.

For example, a flotation unit which has a feed consisting of pyrite (Py) and quartz (Qz) produces a pyrite rich concentrate and a quartz rich tailing (Fig. 4.3):

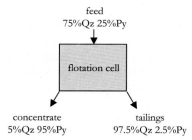

feed
75%Qz 25%Py

flotation cell

concentrate          tailings
5%Qz 95%Py       97.5%Qz 2.5%Py

**Figure 4.3** Typical mass balance for single unit operation.

For this system, the overall mass flow balances for the minerals are simply given by:

$$L_{\text{QZ,in}} = L_{\text{QZ,tails}} + L_{\text{QZ,conc}}$$

$$L_{\text{PY,in}} = L_{\text{PY,tails}} + L_{\text{PY,conc}}$$

in all cases the flow is the product of the flow and concentration:
i.e. $L_{\text{X,stream}} = L_{\text{stream}} \times C_{\text{X,stream}}$

so the above equations can be rewritten as:

$$L_{\text{in}} \times C_{\text{QZ,in}} = L_{\text{tails}} \times C_{\text{QZ,tails}} + L_{\text{conc}} \times C_{\text{QZ,conc}}$$

$$L_{\text{in}} \times C_{\text{PY,in}} = L_{\text{tails}} \times C_{\text{PY,tails}} + L_{\text{conc}} \times C_{\text{PY,conc}}$$

which can be reduced further by substituting the concentrations given

$$L_{in} \times 0.75 = L_{tails} \times 0.975 + L_{conc} \times 0.05$$

$$L_{in} \times 0.25 = L_{tails} \times 0.025 + L_{conc} \times 0.95$$

There are now three unknowns and two equations which are insoluble, knowing only one of the mass flows allows a solution to be determined. However, if the plant is operating at **steady state,** the total mass flow in is equal to the total mass flow out, that is

$$L_{in} = L_{tails} + L_{conc}$$

giving a third equation, making the system soluble as shown below.

$$L_{in} = L_{tails} + L_{conc}$$

substituting gives

$$(L_{tails} + L_{conc}) \times 0.75 = L_{tails} \times 0.975 + L_{conc} \times 0.05$$
$$(L_{tails} + L_{conc}) \times 0.25 = L_{tails} \times 0.025 + L_{conc} \times 0.95$$

which expanding and subtracting gives the following:

$$0.70 L_{conc} = 0.225 \, L_{tails}$$
$$\text{so } L_{conc} = 0.225/0.70 \, L_{tails}$$
$$= 0.321 \, L_{tails}$$

For a complete solution, one mass needs to be specified.

## 4.1.5 Hydrometallurgical plants

Unlike the comparatively simple mineral processing plants which separate phases, hydrometallurgical plants also separate elements by chemical reactions.

In a hydrometallurgical operation, one or more of the following occurs:

○ Phases remain unchanged
  ❑ unreactive minerals
○ Phases disappear
  ❑ dissolution
  ❑ gas consumption
○ Phases form
  ❑ precipitation
  ❑ gas evolution
  ❑ electrowinning

○ Oxidation states of elements change
  ❏ oxidation
  ❏ reduction
○ Ions transfer between phases
  ❏ solvent extraction
  ❏ ion exchange
  ❏ adsorption
○ Enthalpy changes
  ❏ endothermic reactions
  ❏ exothermic reactions
  ❏ temperature of system changes

A schematic for a general hydrometallurgical plant is shown in Fig. 4.4.

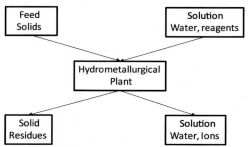

**Figure 4.4** Process streams for a hydrometallurgical plant.

Clearly, this is a far more complex scenario than a mineral processing operation. Instead of balancing only phases and water, far more has to be balanced to give the overview of the plant:

◆ water
◆ individual component minerals, for example quartz, sphalerite, gold
◆ elements, for example nickel, copper, sulphur
◆ dissolved ions, for example $Cu^{2+}$, $Cl^-$, $H^+$, $OH^-$, $Fe^{3+}$, $Fe^{2+}$, $SO_4^{2-}$
◆ electrical charge (electrons cannot exist in isolation)
◆ thermal energy

Hydrometallurgical plants produce chemical changes to some or all of the feed material. For a quartz–pyrite feed, the quartz would remain unreacted, but the pyrite could be oxidised to form dissolved iron ions, dissolved sulphate ions, solid sulphur, solid iron oxide and water as shown in Fig. 4.5.

Thus, simply balancing the mineral flow is only applicable for the quartz which remains unchanged. The pyrite cannot be balanced by the sum of all solid and liquid flows containing iron and/or sulphur as there may be different flows for sulphur

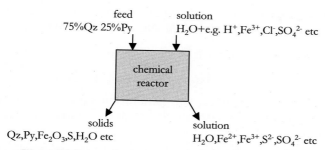

**Figure 4.5** Process streams for a pyrite leaching process.

and iron out of the system but must be separated into separate balances for iron and sulphur.

$$L_{quartz,in} = L_{quartz,out}$$

$$L_{Fe,py} = L_{Fe,aq} + L_{Fe,solids}$$

$$L_{S,py} = L_{S,aq} + L_{S,solids}$$

Although the feed ratio of the iron and sulphur is fixed at 1:2, the ratios in the solids and solution will not necessarily be the same. In most cases, they will be substantially different.

In this instance, if there are no other sources of sulphur or iron, then a further constraint can be made on the basis of the known stoichiometry of pyrite, $FeS_2$:

$$L_{S,py} = 2L_{Fe,py}$$

In all plants, water is a very important phase and is typically by far the majority phase in hydrometallurgical systems and the water balance is of critical importance to the plant:

$$L_{water,in} = L_{water,aq} + L_{water,solids} + L_{water,gas}$$

Both water and hydrogen ions are commonly involved in chemical reactions as either reactant or product and are either consumed or produced in reactions such as follows:

| | |
|---|---|
| At the anode in electrowinning: | $2H_2O = O_2 + 4H^+ + 4e^-$ |
| During pH adjustment with lime: | $Ca(OH)_2 + 2H^+ = 2H_2O + Ca^{2+}$ |
| Oxidation of pyrite: | $2FeS_2 + H_2O + 7.5O_2 = 2Fe^{3+} + 4SO_4^{2-} + 2H^+$ |

Therefore, the water and proton balances are not those given above and must be adjusted for the chemical reaction:

$$L_{water,in} = L_{water,aq} + L_{water,solids} + L_{water,rxn} + L_{water,gas}$$

$$L_{H+,in} = L_{H+,aq} + L_{H+,solids} - L_{H+,rxn}$$

For a reactive system, the generation and consumption of components must be accounted for, and a general balance can be written:

accumulation = in − out + generation − consumption

All of these balances must be considered independently, although some of them may be interdependent.

The concentration, or grade, of each stream is important since this enables the determination of the destination of each element present in the feed material. Again, for every component or element, the mass fed into the plant must be equal to the mass coming out.

Using a simple example of a copper concentrator as shown in Fig. 4.6.

For each stream, the mass of the component is simply its concentration, C, within that stream multiplied by the mass of the stream:

$$M_{X,feed} = C_{X,feed} \cdot m_{feed}$$

Thus, the overall mass balance for X can be written as follows:

$$C_{X,feed} \cdot m_{feed} = C_{X,conc} \cdot m_{conc} + C_{X,tail} \cdot m_{tail}$$

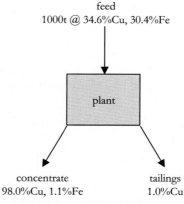

feed
1000t @ 34.6%Cu, 30.4%Fe

plant

concentrate
98.0%Cu, 1.1%Fe

tailings
1.0%Cu

**Figure 4.6** Process stream composition for a copper concentrator.

For each stream either into, or out from, a unit operation there is a concentration times mass term to be added to the above equation.

The mass balance for copper can be written:

$$C_{Cu,feed} m_{feed} = C_{Cu,conc} m_{conc} + C_{Cu,tail} m_{tail}$$

$$34.6 \times 1000 = 98.0 \times m_{conc} + 1.0 \times m_{tail}$$

similarly, for iron:

$$30.4 \times 1000 = 1.1 \times m_{conc} + C_{Fe,tail} \times m_{tail}$$

From these two simultaneous equations and the overall mass balance equation

$$1000 = m_{conc} + m_{tail}$$

it is possible to solve for the three unknowns, $m_{conc}$, $m_{tail}$ and $C_{Fe,tail}$ as shown below.

Solving the simultaneous equations...

$m_{conc} = 1000 - m_{tail}$

substituting this into the copper mass balance

$34.6 \times 1000 = 98.0 \times (1000 - m_{tail}) + 1.0 \times m_{tail}$

and rearranging gives

$$m_{tail} = (98.0 \times 1000 - 34.6 \times 1000)/(98.0 - 1.0)$$

so... $m_{tail} = 653.6$ tonnes and $m_{conc} = 346.4$

these values are then substituted into the iron mass balance

$30.4 \times 1000 = 1.1 \times 346.4 + C_{Fe,tail} \times 653.6$

which, after rearrangement, gives

$$C_{Fe,tail} = (30.4 \times 1000 - 1.1 \times 346.4)/653.6$$

$C_{Fe,tail} = 45.9\%$

checking the copper masses

input $1000 \times 34.6 = 34600$

output $346.4 \times 98.0 + 653.6 \times 1.0 = 33947 + 654 = 34601$

and the iron masses

input $1000 \times 30.4 = 30400$

output $346.4 \times 1.1 + 653.6 \times 45.9 = 381 + 30000 = 30381$

NOTE: the copper output is apparently slightly less (0.06%) than that input. This is due to smaller errors in analysis accumulating during the calculation. The apparent high precision for iron is because the tails analysis was calculated effectively removing analytical errors in the feed and concentrate.

In general, the mass flows will be constantly monitored, but only a limited range of concentrations will be measured on the critical streams. Thus, in the above example, the feed and concentrate may be analysed but the tails may only be analysed for copper.

Using the above information there is an important plant factor which can be determined, namely, the **recovery**.

Recovery is defined as the mass of valuable material reporting to the concentrate or valuable product as a fraction of the total fed into the plant. For the above example

$$\% \text{ Recovery (Cu)} = \frac{m_{Cu,feed}}{m_{Cu,tail}} \times 100$$

or, after substituting the masses and concentrations

$$\%\text{Recovery(Cu)} = C_{Cu,conc}m_{conc}/C_{Cu,feed}m_{feed}$$

Calculating the copper and iron recoveries into the concentrate

$$\%\text{Cu recovery} = 98.0 \times 346.4/(34.6 \times 1000) \times 100$$
$$= 33947/34600 \times 100 = 98.1 \%$$
$$\%\text{Fe recovery} = 1.1 \times 346.4/(30.4 \times 1000) \times 100$$
$$= 381/30400 \times 100 = 1.3 \%$$

This general technique for balancing material within a plant is applicable to all unit processes and plants. However, it is rare for material balances to be calculated by hand due to the complexity of the systems involved, and therefore, computers are far more efficient at solving the complex series of simultaneous equations where a complete plant is being balanced. There are a number of flow sheeting packages which are used in conjunction with experimental data to design and optimise plants.

## 4.2 Energy balances

Unlike the material balance, the energy balance may involve the release or consumption of energy due to reaction or phase transitions within the system.

We need to know the energetics of the plant in order to optimise the

■ Heat generation
  ❑ steam or refrigeration?

- Heat exchangers
  - sizes, locations, etc.
  Reactions which give out energy are called **exothermic** whilst those which absorb energy are called **endothermic**.
  Examples of exothermic processes are
- combustion of coal
- diluting concentrated sulphuric acid
- sulphur oxidation
- water freezing
- phases crystallising from solution
- bacterial oxidation of metal sulphide ore
  endothermic reactions include
- ice melting
- water boiling
- electroreduction of nickel ions
- dissolution of potassium nitrate
- electrogeneration of chlorine
  To determine whether a reaction is exothermic or endothermic, it is necessary to examine the enthalpies of the products and reactants.

$$\Delta H_{reaction} = \Delta H_{products} - \Delta H_{reactants}$$

If the products have a lower enthalpy than the reactants, then energy is released and $\Delta H_{reaction}$ is negative. On the other hand, if the products have a higher enthalpy than the reactants heat is required to complete the reaction and $\Delta H_{reaction}$ is positive.

This relationship holds for all reactions at all temperatures with adjustments made for the change in enthalpy with temperature as outlined in Chapter 2.

For example

**(1)** The burning of sulphur in oxygen

$$S + O_2 = SO_2$$

$\Delta H_{SO_2} = -297 \text{ kJ mol}^{-1}$ (data given in Appendix 1 of Chapter 2)
By definition, the enthalpies of formation of elements at STP are 0 kJ mol$^{-1}$

$$\Delta H_{reaction} = -297 - (0+0) = -297 \text{ kJ mol}^{-1}$$

**(2)** Calcining of $Ca(OH)_2$

$$Ca(OH)_2 = CaO + H_2O$$

$$\Delta H_{Ca(OH)_2} = -986.9 \text{ kJmol}^{-1}, \Delta H_{CaO} = -634.7 \text{ kJmol}^{-1},$$

$$\Delta H_{H_2O} = -286.1 \text{ kJmol}^{-1}$$

$$\Delta H_{reaction} = (-634.7 - 286.1) - (-986.9) = +66.1 \text{ kJmol}^{-1}$$

**(3)** Electrodeposition of cobalt

$$Co^{2+} + 2e^- = Co$$

By definition, the enthalpies of formation of elements and electrons at STP is 0 kJ mol$^{-1}$

$$\Delta H_{reaction} = -\Delta H_{Co2+} = -58.158 \text{ kJmol}^{-1}$$

**(4)** electrooxidation of chloride ions:

$$2Cl^- = Cl_2 + 2e^-$$

$$\Delta H_{reaction} = -2\Delta H_{Cl-} = -2(-167.080) = 334.160 \text{ kJmol}^{-1}Cl_2$$

Note that in this case, we must specify 1 mole $Cl_2$ as the product.

Using the data in the Appendices to Chapter 2 to determine whether the following reactions are exothermic or endothermic at 25°C:

$$C + O_2 = CO$$

$$NaCl = Na^+ + Cl^-$$

$$Pb^{2+} + 2e^- = Pb$$

$$CuFeS_2 + O_2 = Cu^{2+} + Fe_2O_3 + S + SO_4^{2-}$$

$$FeS_2 + H_2O + Fe^{3+} = Fe^{2+} + SO_4^{2-} + H^+$$

$$H_2SO_4 = H^+ + HSO_4^-$$

$$NiO + HCl = H_2O + Ni^{2+} + Cl^-$$

$$Au + 2CN^- = Au(CN)_2^- + e^-$$

In almost all metallurgical processes, there is a change in the heat contained within the system due to reaction. It is important to maintain the process at the optimum temperature to ensure that operational and control problems do not occur. In a manner similar to a mass balance for a reacting system, we need to ensure that there is a corresponding energy balance.

Thus, in general terms we can depict a typical system as shown in Fig. 4.7.

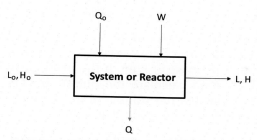

**Figure 4.7** Components of an energy balance.

The energy balance for the system is
Energy in = Energy out

$$L_o.H_o + Q_o + W = L.H + Q$$

in which, $Q_o$ is the heat added to the reactor, $L_o$ and $L$ are the flow-rates of species with enthalpies $H_o$ and $H$, respectively, entering and leaving the system while $Q$ is the heat lost from the reactor. $W$ is the work done on the reactor.

Flow rate $= 10 \text{ kgs}^{-1}$
$= 10 \times 1000/18 = 555.6 \text{ mols}^{-1}$

Assume that $C_p^o$ is constant between 25 and 60°C, in which case,
$H = H_f^o + (60{-}25)\,(75.3)/100 \text{ kJ}$
Substituting in the heat balance,
$555.6(-286.1) + 1500 = 555.6(-286.1) + 555.6(35)(75.3)/1000 + Q$
$Q = 35.7 \text{ kJ s}^{-1} = 35.7 \text{ kW}$
i.e a loss of 35.7(100)/1500 or 2.4% of power input.

Water flows into a closed tank at 10 kg s$^{-1}$ and 25°C where it is heated by an electric element rated at 1500 kW. If the temperature leaving the tank is 60°C, estimate the energy losses through the walls of the tank.
($H_f^o = -286.1 \text{ kJ mol}^{-1}$, $C_p^o = 75.3 \text{ J mol}^{-1} \text{ K}^{-1}$ for water)

## 4.3 Examples

A number of examples of both mass and energy balance calculations are given in the following sections. A spreadsheet such as Excel is convenient to solve mass and energy balance problems.

### 4.3.1 A mixer

A gold plant uses a 5 % solution of $Ca(OH)_2$ (Lime) to maintain the pH at 9.6. However, the $Ca(OH)_2$ is supplied as a slurry containing 50 % by mass of lime. If the plant consumes 5 t d$^{-1}$ of the stock solution calculate the pump size needed for the dilution water flow and the volume of the mixing tank required to ensure a residence time of 1 h (Fig. 4.8).

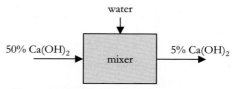

**Figure 4.8** Streams in a simple mixing unit.

$Ca(OH)_2$ mass balance,

$$M_{Ca(OH)_2,In} = M_{Ca(OH)_2,Out}$$

Water mass balance

$$M_{H_2O,In} + M_{H_2O,Add} = M_{H_2O,Out}$$

For 5 t d$^{-1}$ of the 50% $Ca(OH)_2$ slurry.
$5 \times 0.50 = 2.5$ tonnes $Ca(OH)_2$ and 2.5 tonnes water.

The final solution also contains 2.5 tonnes $Ca(OH)_2$ which is 5% of the total mass 2.5/0.05 =50 tonnes making the water 47.5 tonnes.

The water balance is

$$2.5 + M_{H_2O,Add} = 47.5$$

Thus, the final solution requires the addition of 45 tonnes water per day.

For water the hourly flowrate is given by:
45 tpd of water is 45/24 t h$^{-1}$ = 1.875 t h$^{-1}$ water.

Converting to volume 1.875 t h$^{-1}$ = 1.875 t h$^{-1}$/0.99657 t m$^{-3}$ = 1.881 m$^3$ h$^{-1}$

The mixer needs to allow 50/24 = 2.083 tonnes of slurry to be mixed per hour, assume that the 5 % solution can be treated as a physical mixture of the two components:

$$\text{i.e. } V_{Out} = 0.95M_{H2O}/\rho_{H2O} + 0.05\ M_{Ca(OH)_2}/\rho_{Ca(OH)_2}$$
$$= 0.95(45/0.99657) + 0.05(5/2.24)$$
$$= 42.897 + 0.112$$
$$= 43.009 \text{ m}^3 \left(\text{giving a density of } 50/43 = 1.16 \text{ tm}^{-3}\right)$$

so, each hour a volume of $2.083/1.16 = 1.80 \text{ m}^3$ needs to be mixed.

In practice, the tank would probably be $2 \text{ m}^3$ to allow for some variation in flowrate and to allow some recirculation of solution which may aid mixing.

## 4.3.2 A simple batch process

A batch reactor operates to dissolve zircon using sodium hydroxide, the required amounts of each component are fed into a vessel which is then heated to 600°C where the reaction is complete after one hour and then cooled to room temperature. The sodium silicate formed is insoluble in the acid used whilst the sodium zirconate is soluble. The total cycle time is 4 h.

Determine the amount of sodium hydroxide to be added if the reactor can hold 500 kg of zircon at a time. Also, what volume of steam needs to handled, assume that the reaction occurs at 600°C. Approximately what mass loss can be expected from the product solids after dissolution in acid?

Schematically this is shown in Fig. 4.9.

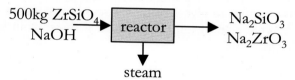

**Figure 4.9** Streams in a simple batch reactor.

The overall reaction is given by:

$$ZrSiO_4 + 4NaOH = Na_2ZrO_3 + Na_2SiO_3 + 2H_2O$$

In each batch, there are $500/0.18303 = 2731.8$ moles zircon, and therefore $4 \times 2731.8 = 10927$ moles NaOH, are required

$$= 10927 \times 39.997 = 437047 \text{ g} = 437.05 \text{ kg NaOH}$$

From the reaction stoichiometry, a total of $2 \times 2731.8 = 5463.6$ moles of water are evolved, or $5463.6 \times 18.015 = 98427 \text{ g}$ $= 98.427$ kg water.

Using Boyles gas law $V_1/T_1 = V_2/T_2$

At $T_1 = 273.15$ K, $V_1 = 0.022414 \text{ m}^3 \text{ mol}^{-1}$ (molar volume of an ideal gas at STP), therefore at $T_2 = 873.15$ K

$$V_2 = 0.022414/273.15 \times 873.15 = 0.071648 \text{m}^3\text{mol}^{-1} \text{ at } 600°C$$

Therefore, for 5463.6 moles $V = 0.071648 \times 5463.6 = \mathit{391.46\ m^3}$ *of steam.*

For each mole of zircon in, one mole each of $Na_2ZrO_3$ and $Na_2SiO_3$ are formed.

Thus, there are 2731.8 moles of $Na_2ZrO_3$ and 2731.8 moles of $Na_2SiO_3$ formed.

Or $M_{Na2ZrO3} = 2731.8 \times 185.198 = 505924g = 505.9$ kg $Na_2ZrO_3$
and $M_{Na2SiO3} = 2731.8 \times 122.063 = 333452g = 333.5$ kg $Na_2SiO_3$
%   $Na_2ZrO_3 = 505.9/(505.9 + 333.5) \times 100 = 505.9/839.4 \times 100$
$= 60.3\%$ *dissolved.*

Determine the amount of energy needed to heat the reactants to 600°C?

Is the reaction exothermic or endothermic at 600°C?

If all of this energy is used to heat the solid products what is the final temperature of the solids?

### 4.3.3  A simple flow process

In the above example, the acid leaching is to be carried out continuously at pH 0 in a small tank, what flowrate of acid and solids are required to give an outlet zirconium concentration of $100$ g $L^{-1}$. What size of tank is required if the dissolution takes 15 mins? (Fig. 4.10).

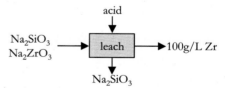

**Figure 4.10** Streams in a continuous leach process.

Schematically.

$Na_2ZrO_3$ is $91.22/(2 \times 22.99 + 91.22 + 3 \times 16.00) \times 100$

$= 49.25$ wt%Zr

Thus, if $505.9$ kg $Na_2ZrO_3$ dissolves there is a total of $505.9 \times 0.4925 = 249.2$ kg Zr dissolved.

For $100$ g Zr $L^{-1}$ a total acid volume of $249.2/(100/1000) = 2492$ L is required.

The cycle time is 4 h for all dissolution to occur; therefore, the solids flow is $839.4/4 = 209.9$ kg $h^{-1}$

and the acid flow is $2492/4 = 623$ L $h^{-1}$

Assume the density of the solids is $2.5$ kg $L^{-1}$ (no data available for $Na_2ZrO_3$, but $2.4$ kg $L^{-1}$ for $Na_2SiO_3$)

Volume of solids $= 209.9/2.5 = 84.0 \text{ L h}^{-1}$
Thus, total volumetric flow rate $= 623 + 84.0 = 707 \text{ L h}^{-1}$
Assume that the mean residence time is 15mins.
Thus, the volume of tank required is $707 \times 15/60 = 177 \text{ L}$.
*In practice the tank would be 200 L to allow a slightly longer residence time, thereby ensuring that the dissolution of the zirconium phase was complete and handling any flow surges.*

## 4.3.4 A batch process with heat transfer

An electrowinning cell runs at 25°C at pH 0 and contains 200 g $\text{L}^{-1}$ of cobalt(II) chloride which is electrowon onto a cobalt cathode, the anode generates chlorine. If the volume of the cell is 10 L, what volume of chlorine is formed, how much heat needs to be added to maintain the cell at 25°C, how much charge needs to be passed to reduce the cobalt content of the cell to 50 ppm and how much NaCl needs to be added to maintain the chloride at $>0.1$ M?

The unit is shown schematically in Fig. 4.11.
There are two electrochemical reactions...

$$\text{At the cathode: } Co^{2+} + 2e^- = Co$$

$$\text{At the anode: } 2Cl^- = Cl_2 + 2e^-$$

200 g/L $CoCl_2$ contains $200 \times 58.933/(58.933 + 2 \times 35.453) = 90.778$ g Co and 109.222 g $Cl^-$

Start: $Co = 10 \text{ L} \times 90.778 \text{ g/L} = 907.78 \text{ g} = 907.78/58.933 = 15.404$ moles

$$Cl = 2 \times \text{cobalt} = 15.404 \times 2 = 30.808 \text{ moles } Cl^-$$

End: 50 ppm $Co = 50 \times 10^{-3}\text{g/L} = 50 \times 10^{-3} \times 100 = 5.0 \text{ g} = 5.0/58.933 = 0.0848$ moles Co

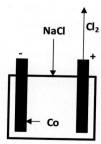

**Figure 4.11** Components of the mass balance for an electrolytic cell.

Thus, 15.404−0.0848 = 15.319 moles of Co are electrowon.

Thus, 15.319 × 2 = 30.638 moles of electrons are passed to win the Co.

An equal amount of charge is passed to evolve chlorine, but one chlorine is evolved per two moles of electrons,

therefore, 30.638/2 = 15.319 moles of chlorine are evolved,

*from  Boyles  law    15.319 × 0.02241383 × 298.15/273.15 = 0.3748 m³ of chlorine is evolved*

and 30.638 moles of chloride are consumed leaving 30.808− 30.638 = 0.17 moles in solution

but  0.17  moles = 0.17/10 = 0.017 M,  which  is  0.1−0.017 = 0.083 M below that required

that is 0.083 × 10 = 0.83 moles of Cl need to be added to maintain the concentration.

0.83 moles of chloride are in 0.83 moles of NaCl

*therefore 0.83 × 58.443 = 48.51 g NaCl needs adding to maintain 0.1 M chloride.*

For the cobalt electrowinning, the enthalpy of reaction is $-58.199$ kJ mol$^{-1}$ Co

for chlorine evolution the enthalpy is 334.385 kJ mol$^{-1}$ Cl$_2$

thus, the total enthalpy is $-58.199 + 334.385 = 276.186$ kJ mol$^{-1}$ CoCl$_2$

in this case the total energy consumed is 15.404 × 276.186 = 4254.4 kJ and this amount

must be added to the bath to maintain it at 25°C.

In this and the example in Section 4.3.6, we have assumed that the electrolysis process operates under ideal conditions, that is without any overpotentials at the electrodes and with an electrolyte with zero resistance. This is obviously not the case and we will deal with this shortcoming in Chapter 14.

## 4.3.5 A simple flow process with heat transfer

A bacterial oxidation tank operates at pH 2, the feed is 10 tonnes h$^{-1}$ of an ore containing quartz and 35% pyrite which is completely oxidised. Cooling water is available at 10°C. Determine the volume of air required, the amount of 50% NaOH and the volume of cooling water necessary to maintain the tank operating at 25°C. Ignore all losses from the system.

The process streams are shown in Fig. 4.12.

Pyrite feed rate: 10 × 35/100 = 3.5 tonnes h$^{-1}$

The equation for the complete oxidation of pyrite is given by:

$$2FeS_2 + 7.5O_2 + H_2O = 2Fe^{3+} + 4SO_4^{2-} + 2H^+$$

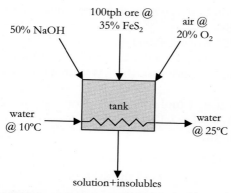

**Figure 4.12** Process streams for a continuous leach process.

Thus, for each mol of pyrite, 3.75 moles of oxygen are required and 1 mole of protons is produced.

Thus, in each hour:

Feed: $3.5 \times 10^6$ g $FeS_2/(55.85 + 2*32.06)$ g mol$^{-1}$ $FeS_2$ - = 29174 mol $FeS_2$

this requires $29174 \times 3.75 = 109400$ mol of $O_2$

At 25°C 1 mol of gas occupies 22.414 L mol$^{-1}$ × 298.15 K/ 273.15K = 24.465 L

Therefore, 2676500 L of oxygen are required. Assuming air is 20 % oxygen.

The volume of air required is $100/20 \times 2676500 = 13382500$ L. = 13382.5 m$^3$

To maintain pH requires 29174 moles of sodium hydroxide.

29174 moles = $29174 \times (22.99 + 16.00 + 1.01)$ g NaOH = 1166960 g.

But the NaOH is a 50% solution, so

the mass of the solution is $100/50 \times 1166960$ g = 2333920 g ($\sim 2.33$ tonnes)

From the oxidation equation the enthalpy released can be calculated as

$$\Delta H = \left(2\Delta H_{Fe_{3+}} + 4\Delta H_{SO_4^{2-}} + 2\Delta H_{H+}\right)$$

$$-(2\Delta H_{FeS_2} + 7.5\Delta H_{O_2} + \Delta H_{H_2O})$$

$$= (2x - 48.5 + 4x - 909.4 + 2x0.0)$$

$$-(2x - 171.7 + 7.5x0.0 - 286.1)$$

$$= -3105.1 \text{ kJ per 2 mole } FeS_2 \text{or} = -1552.6 \text{ kJ mol}^{-1} FeS_2$$

Thus, the total energy released every hour by the oxidation of 29174 mol $FeS_2$
29174 × 1552.6 = 45295600 kJ.
The heat capacity of water at 25°C is 75.3 J mol$^{-1}$ K$^{-1}$.
For simplicity, assume this is constant over the range 10–25°C.
Thus to raise 1 mol of water from 10 to 25°C requires 75.3 × 15 = 1129.5 J = 1.13 kJ.
So, every hour

$$45295600/1.13 = 40085000 \text{ moles of water are needed to}$$
$$\text{cool the system}$$
$$= 40085000 \times 0.018 = 721530 \text{ kg}$$

Repeat the calculation with the tank at 35°C.
Comment on the reasons for the different answers.

## 4.3.6 A recycle process

An acid leach for a pure copper sulphide is run at pH 0 using hydrochloric acid and chlorine, the copper is then electrowon and the acid and chlorine recycled to the leach tank. How much chlorine, NaOH, and water need to be added to maintain the system running if 0.1 t h$^{-1}$ $Cu_2S$ is oxidised?
The process flow diagram is shown in Fig. 4.13.

### 4.3.6.1 Mass balance

In the leach:

$$Cu_2S + 5Cl_2 + 4H_2O = 2Cu^{2+} + SO_4^{2-} + 8H^+ + 10Cl^-$$

Use a basis of one hour...
Every hour 100 kg = 100/159.152 = 0.6283 kmol $Cu_2S$ is oxidised
requiring 5 × 0.62833 = 3.1416 kmol $Cl_2$ and 4 × 0.6283 = 2.5133 kmol $H_2O$
Leach produces 2 × 0.6283 = 1.2566 kmol $Cu^{2+}$, 0.6283 kmol $SO_4^{2-}$,
8 × 0.6283 = 5.0266 kmol $H^+$ and 10 × 0.6283 = 6.2833 kmol $Cl^-$
For constant pH, 5.0226 kmol $OH^-$ needs to be added as NaOH
*5.0226 kmol NaOH is 5.0226 × 39.997 = 200.89 kg NaOH*

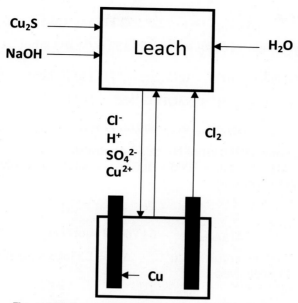

**Figure 4.13** Process flows for two coupled operations.

The neutralisation reaction:
$$NaOH + H^+ = H_2O + Na^+$$
produces 5.0226 kmol of water per hour.

The water balance is therefore $5.0226 - 2.5133 = 2.5093$ kmol water gained

*Thus for constant volume $2.50932$ kmol $= 2.5093 \times 18.015$ $= 45.205$ kg water needs to be removed every hour*

In electrowinning:

$$Cu^{2+} + 2e^- = Cu$$

$$2Cl^- = Cl_2 + 2e^-$$

Each hour, 1.2566 kmol $Cu^{2+}$ is electrowon using $2 \times 1.2566 = 2.5133$ kmol electrons

and this charge evolves $2.5133/2 = 1.25666$ kmol $Cl_2$.

Thus, the net chlorine requirement is $3.14165 - 1.25666 = 1.88499$ kmol

$$1.88499 \text{ kmol} = 1.88499 \times 70.906 = 133.66 \text{ kg } Cl_2$$

### 4.3.6.2 Energy balance

In the leach:

$$Cu_2S + 5Cl_2 + 4H_2O = 2Cu^{2+} + SO_4^{2-} + 8H^+ + 10Cl^-$$

$$\Delta H_{reaction} = -1227.292 \text{ kJ mol}^{-1} Cu_2S$$

Every hour $100 \text{ kg} = 100/159.152 = 0.6283 \text{ kmol}$ $Cu_2S$ is oxidised

and $0.62833 \times 1000 \times 1227.292 = 77114 \text{ kJ}$ of heat is evolved

$$H^+ + NaOH = Na^+ + H_2O$$

$$\Delta H_{reaction} = -100.06 \text{ kJ mol}^{-1} H^+$$

each hour $5.0266 \text{ kmol } H^+$ need neutralising:

Thus $\Delta H = 5.0266 \times 1000 \times 100.061 = 502971 \text{ kJ}$ evolved

In electrowinning:

$$Cu^{2+} + 2e^- = Cu$$

$$\Delta H_{reaction} = -64.768 \text{ kJ mol}^{-1} Cu$$

for $1.2566 \text{ kmol}$ $Cu^{2+}$, $\Delta H = 1.25666 \times 1000 \times 64.768$ $= 81391 \text{ kJ}$ is evolved

$$2Cl^- = Cl_2 + 2e^-$$

$$\Delta H_{reaction} = 334.385 \text{ kJ mol}^{-1} Cl_2$$

for $1.2566 \text{ kmol}$ $Cl_2$, $\Delta H = 1.25666 \times 1000 \times 334.385$ $= 420208 \text{ kJ}$ is consumed.

*Thus the overall balance* $= 771144 + 502971 + 81391 - 420208$ $= 935296 \text{ kJ is evolved}$

As a check, the overall reaction is given by:

$$Cu_2S + 3Cl_2 + 8NaOH = 2Cu + 4H_2O + 8Na^+ + 6Cl^- + SO_4^{2-}$$

For which $\Delta H_{reaction} = -1488.543 \text{ kJ mol}^{-1} Cu_2S$

Thus for $628.33 \text{ mol } Cu_2S$, $\Delta H = 628.33 \times 1488.543 = 935296 \text{ kJ}$ is evolved.

## 4.3.7 A process involving a bleed stream

In the above example, the chloride and sulphate in the solution increase with time as these are not consumed but simply cycle around. Unfortunately, if the sulphate concentration in the electrowinning feed solution rises above $10 \text{ g L}^{-1}$, precipitation occurs. Thus, it is necessary to have a bleed stream where a volume of solution is diverted to a tank where $Ca(OH)_2$ is added to precipitate $CaSO_4.2H_2O$ leaving $1\text{g L}^{-1}$ sulphate in solution. If the total volume of solution in the leach tank is $8 \text{ m}^3$ determine the volume of solution in the bleed stream and the mass of $Ca(OH)_2$ required.

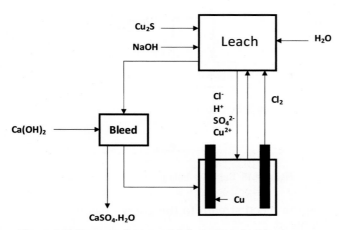

**Figure 4.14** Process diagram with inclusion of a bleed stream.

The process flow diagram is shown in Fig. 4.14.

In the bleed:

$$Ca(OH)_2 + SO_4^{2-} + 2H^+ = CaSO_4.2H_2O$$

From the last example, every hour $0.628$ kmol $SO_4^{2-}$ and $8 \times 0.628 = 5.026$ kmol $H^+$ are produced.

Thus, to remain at $10$ g $L^{-1}$, $0.628$ kmol $SO_4^{2-}$ needs to be removed using.

$0.628$ kmol $= 0.628 \times 74.095 = 46.556$ kg $Ca(OH)_2$

HOWEVER, this process also consumes $2 \times 0.628 = 1.256$ kmol $H^+$ so less NaOH is required to maintain pH in the leach tank.

For constant pH, $5.022-1.256 = 3.765$ kmol $OH^-$ needs to be added as NaOH.

*3.765 kmol NaOH is 3.568 × 39.997 = 150.63 kg NaOH.*

Assume that the volume of solution treated is $8$ m$^3$ per hour, the sulphate increase is $0.628$ kmol $= 0.628 \times 96.058 = 60.356$ kg $SO_4^{2-}$

$= 7.545$ g $L^{-1}$ which has to be removed.

Examining the solution sulphate balance in the bleed in more detail as shown in Fig. 4.15.

Concentrating on the mass balance around the mixing point before electrowinning

for sulphate $V_{electrowin} \times 10 = V_{bleed} \times 1.0 + V_{non-bleed} \times 17.545$

for solution $V_{electrowin} = V_{bleed} + V_{non-bleed}$

thus with $8$ m$^3$ entering electrowinning and solving the simultaneous equations.

$V_{bleed} = 3.648$ m$^3$

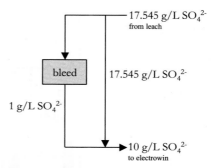

**Figure 4.15** Sulphate balance in the bleed stream.

Adding the bleed stream to the energy balance. In the bleed

$$Ca(OH)_2 + SO_4^{2-} + 2H^+ = CaSO_4.2H_2O$$

$$\Delta H_{reaction} = -127.152 \text{ kJ mol}^{-1} Ca(OH)_2$$

for 0.62833 kmol $SO_4^{2-}$  $\Delta H = 0.62833 \times 1000 \times 127.152 = 79893$ kJ is evolved.

However, after bleed, the amount of $H^+$ needing to be neutralised is decreased to 3.76598 kmol (as the gypsum precipitate also removes $H^+$)

$$\Delta N_{eutralisation} = 3.76598 \times 1000 \times 100.061 = 376827 \text{ kJ is evolved}$$

so the new heat balance is

*771144 + 376827 + 81391 + 79893   −   420208 = 889047 kJ  is evolved.*

To check this, the overall reaction is....

$$Cu_2S + 3Cl_2 + 6NaOH + Ca(OH)_2$$

$$= 2Cu + CaSO_4.2H_2O + 2H_2O + 6Na^+ + 6Cl^-$$

$$\Delta H_{reaction} = -1415.573 \text{ kJ mol}^{-1} Cu_2S$$

thus, for 628.33 mol $Cu_2S$, $\Delta H = 628.33 \times 1415.573 = 889447$ kJ is evolved.

## 4.3.8 A solvent extraction process

A uranium plant operates the solvent extraction plant pictured shown in Fig. 4.16, what ratio of organic to aqueous is required in the mixer-settler and what volume of strip solution is required? If an annual production of 1000 t of yellow cake is required, what is the solvent inventory if the mixing, settling, and extraction takes 10 min?

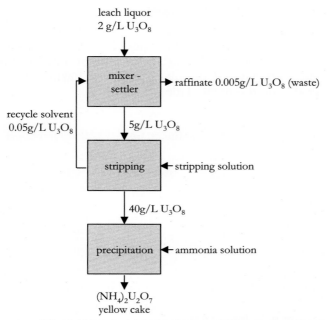

**Figure 4.16** Streams in a typical solvent extraction process.

Assume a feed of 1 L (unit time)$^{-1}$ for simplicity.

From each litre of feed $U_3O_8$ extracted is $2-0.005 = 1.995$ g.

Since the solvent circulates with $0.05$ g $L^{-1}$ it gains $5-0.05 = 4.95$ g $L^{-1}$

to achieve 5 g $L^{-1}$ requires $1.995/4.95 = 0.403$ L of solvent.

The strip process removes $5-0.05 = 4.95$ g $U_3O_8$ $L^{-1}$ of solvent.

Since the exit solution contains 40 g $L^{-1}$ its volume is $4.95/40 = 0.124$ L.

Thus for each litre of feed solution, $0.124 \times 0.403 = 0.050$ L of strip solution is required.

1000 t a$^{-1}$ of yellow cake $= 1000 \times (2 \times 238.029)/624.130 = 762.75$ t a$^{-1}$ of U.

40 g L$^{-1}$ $U_3O_8$ $= 40 \times (3 \times 238.029)/842.082 = 33.92$ g U L$^{-1}$

Thus, $762.75 \times 10^6/33.92 = 22486700$ L year$^{-1}$

or $22486700/365 \times 24 = 2567$ L h$^{-1}$ of concentrate solution is required

this is equivalent to $2567/0.050 = 51340$ L feed solution h$^{-1}$

each litre of feed requires 0.403 L of solvent, so each hour $51340 \times 0.403 = 20690$ L of solvent are needed

each hour a total of $60/10 = 6$ recirculation's of solvent are made

so the solvent inventory is $20690/6 = 3448$ L.

## 4.3.9 A carbon-in-pulp plant

A gold plant operates a counter-current CIP system, using a series of five tanks for which the operating data are given in the table below. Determine the carbon to solution ratio and the recovery in each stage. If the plant produces 100000 oz y$^{-1}$, how much gold is on the recirculating carbon?

|  | **Au in solution, ppm** | **Gold on carbon, ppm** |
|---|---|---|
| Feed pulp | 1.89 | — |
| Tank 1 exit | 0.52 | 1710 |
| Tank 2 exit | 0.24 | 684 |
| Tank 3 exit | 0.15 | 342 |
| Tank 4 exit | 0.09 | 171 |
| Tank 5 exit | 0.05 | 137 |
| Carbon feed |  | 45 |

A schematic of the process is shown in Fig. 4.17.

The overall process can be reduced to that shown in Fig. 4.18. Thus, the gold balance is:

$$Au_{aq,in} + Au_{C,in} = Au_{aq,out} + Au_{C,out}$$

$$V_{aq,in} \times C_{Au,aq,in} + M_{C,in} \times C_{Au,C,in}$$
$$= V_{aq,out} \times C_{Au,aq,out} + M_{C,out} \times C_{Au,C,out}$$

The volume of solution in and out are equal, as are the masses of carbon in and out.

So:

$$1.89\, V_{aq} + 45\, M_C = 0.05\, V_{aq} + 1710\, M_C$$

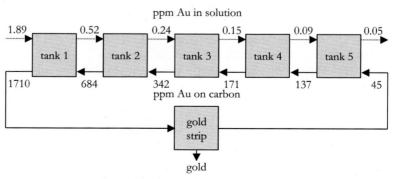

**Figure 4.17** Flowsheet for a CIP process.

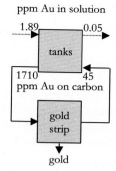

ppm Au in solution

**Figure 4.18** Simplified overall CIP flowsheet.

$1.84 \, V_{aq} = 1665 \, M_C$

thus $M_C/V_{aq} = 1.84/1665 = 0.00111$, or 1.11 g carbon $L^{-1}$

Examining the first tank:

Recovery $= M_{Au,C,out}/M_{Au,aq,in}$

In this case, it is the extra gold on the carbon that is important, so.

Recovery $= (M_{Au,C,out} - M_{Au,C,in})/M_{Au,aq,in}$

Assume a solution volume of 1 L

$= (0.00111 \times 1710 - 0.00111 \times 684)/1 \times 1.89 = 60.26 \%$

For the other tanks:

$R_2 = (0.00111 \times 684 - 0.00111 \times 342)/0.52 = 73.00 \%$

$R_3 = (0.00111 \times 342 - 0.00111 \times 171)/0.24 = 79.09 \%$

$R_4 = (0.00111 \times 171 - 0.00111 \times 137)/0.15 = 25.16 \%$

$R_5 = (0.00111 \times 137 - 0.00111 \times 45)/0.09 = 113.5 \%$

The recoveries in the final two stages are clearly different from those in earlier stages and suggest a problem with the analyses.

From the stepwise recoveries, the overall recovery can be calculated:

$R = (1.89 \times 0.6026 + 0.52 \times 0.7300 + 0.24 \times 0.7909 + 0.15$
$\times 0.2516 + 0.09 \times 1.135)/1.89$

$= (1.140 + 0.380 + 0.190 + 0.038 + 0.102)/1.89$

$= 1.85/1.89 = 97.8 \%$

To check, for each litre of feed, $1.89 - 0.05 = 1.84$ ppm out of 1.89 ppm is not recovered therefore recovery $= 1.84/1.89 = 97.4 \%$

100000 oz = 100000 × 31.10345 = 3110345 g Au (gold plants use Troy ounces)

Thus, each hour 3110345/(365 × 24) = 355.1 g of gold is produced.

Thus, each hour the gold feed to the plant contains 355.1 × 100/97.4 = 364.6 g Au.

The solution contains 1.89 ppm = 0.00189 g L$^{-1}$.

Thus the hourly flowrate is 364.6/0.00189 = 192910 L

which requires 192910 × 1.11 = 214130 g carbon.

The recirculating carbon contains 45 ppm Au = 0.000045 g Au (g carbon)$^{-1}$

Thus the recirculating carbon contains 214130 × 0.000045 = 9.636 g Au

or 9.636/31.10345 = 0.31 oz of gold.

## 4.4 A dose of reality

The preceding examples have been in situations where the system is effectively sealed but these are few and far between in practice. Real systems are somewhat more complex and typically involve using heat capacities to determine how much energy is required to either heat or cool the system to the desired extent. The calculations which result from this more realistic scenario are more complex than the simple examples shown thus far. Indeed, the calculations get ever more complex, especially with systems where the reactions in the system are operated at different temperatures and pressures, for example leaching at room temperature and pressure, followed by gaseous reduction at 200°C under 30 atm pressure. For each movement of material, both mass and heat balances have to be made with each separate change in temperature adding to the complexity. The examples have not considered the inevitable losses in the heat balance due to conduction, convection, and radiation (all of which will be different according to the temperature gradient within the chosen areas of the plant), no process is perfect and any system inevitably heads towards equilibrium with its environment, thereby losing or gaining energy.

Additional complexity comes from the need to recycle reagents such as acid and real plants have multiple recycle streams to maximise the usage of reagents. Real hydrometallurgical systems also have limitations on the actual kinetics of reactions — in an ideal world reactions go to completion rapidly;

in reality they rarely do. The conditions are chosen to maximise the profitability through a combination of factors, for example reducing the capital cost of the plant by installing smaller leach tanks and sacrificing some recovery (many gold cyanidation plants could recover more gold if the leaching residence time was greater, but the extra gold does not cover the cost of increasing the plant size). Clearly, in addition to the mass and heat balances, there is a further financial imperative that the plant must be profitable and this is an additional constraint on the whole system.

As the calculations get increasingly complex, the value of various types of mass and energy balance software becomes important. There are a number of programmes which simultaneously solve mass and heat balances for whole plants at varying levels of complexity. These programmes incorporate a very wide range of parameters not even touched upon in the examples, such as particle size distributions, multiple recycle streams, heat exchangers, and so on. Many also incorporate models of various physical processes and chemical reactors the starting parameters of which can be provided by small-scale experimentation. Using these programmes, simulations approaching reality can be made to "road test" the sensitivity of a proposed or operating plant to various changes in the flowsheet.

## 4.5 Summary

This Chapter has outlined the principles and methods to be used in the analysis of the material and energy balances for hydrometallurgical operations. The approach has been to use example calculations to guide the reader through the methodology. From simple mineral balance problems, we have expanded the calculations to elemental balances that are required as a result of chemical reactions. The methods have been used in simple cases involving only one unit operation to several linked operations including bleed streams.

Many of the examples have included energy balances where necessary, and the reader has been guided through the use of appropriate thermodynamic data to compile such a balance. The assumptions made in such calculations have been pointed out in some cases. Many excellent process engineering software packages are available which enable the engineer to undertake these tasks for complex flowsheets.

# Further reading

Chopey, N.P. (Ed.), 2004. Handbook of Chemical Engineering Calculations, third ed. McGraw-Hill, New York, NY.

Dreisinger, D.B., Dixon, D.G., 2002. In: Taylor, P.R. (Ed.), Hydrometallurgical Process Modelling for Design and Analysis. Part I. Mass and Heat Balances. EPD Congress 2002. TMS, Warrendale, PA.

Mular, A.L., Halbe, D.N., Barrat, D.J., 2002. Mineral Processing Plant Design, Practice and Control, Vols. 1 and 2. SME, Littleton, CO.

Missen, R.W., Mims, C.A., Saville, B.A., 1999. Chemical Reaction Engineering and Kinetics. Wiley and Sons, New York, NY.

Seider, W.D., Seader, J.D., Lewin, D.R., 1999. Process Design Principles. Wiley and Sons, New York, NY.

# Practice problems

**1.** A gold mine has achieved the following production data throughout its lifespan.

| Year | Tonnes | Grade g t$^{-1}$ | kg gold | Residue g t$^{-1}$ |
|------|--------|------------------|---------|--------------------|
| 1991 | 147849 | 6.79 | 965 | 0.29 |
| 1992 | 143489 | 5.53 | 754 | 0.29 |
| 1993 | 147692 | 5.90 | 821 | 0.34 |
| 1994 | 135934 | 6.29 | 865 | 0.32 |
| 1995 | 151224 | 5.54 | 837 | 0.30 |
| 1996 | 155966 | 5.87 | 916 | 0.35 |
| 1997 | 164235 | 5.37 | 881 | 0.31 |
| 1998 | 182849 | 5.29 | 968 | 0.32 |
| 1999 | 165310 | 5.33 | 881 | 0.27 |
| 2000 | 202476 | 5.18 | 1049 | 0.27 |
| 2001 | 204510 | 5.58 | 1085 | 0.27 |

    **(a)** calculate the annual gold recovery from the ore
    **(b)** discuss the results
    **(c)** based on the company data for the residue, how much gold is in the tailings?
    **(d)** based on a mass balance, how much gold is in the tailings?
    **(e)** which do you consider to be more reliable?
    **(f)** your company is interested in reprocessing the tails, but only if the grade is $>0.25$ g t$^{-1}$, what do you recommend?

2. Three similar lead-zinc concentrators operate as follows

|  | Tonnes | %Pb | g t$^{-1}$ Ag | %Zn |
|---|---|---|---|---|
| Feed | 491510 | 12.94 | 229 | 10.24 |
| Pb con | 82487 | 74.74 | 1281.2 | 4.12 |
| Zn con | 85809 | 1.02 | 35.7 | 52.76 |
| Feed | 862684 | 9.0 | 68 | 8.7 |
| Pb con | 96248 | 76.1 | 545 | 3.5 |
| Zn con | 132383 | 1.38 | 25 | 51.6 |
| Feed | 1071780 | 6.9 | 58 | 11.6 |
| Pb con | 91017 | 73.9 | 570 | 4.0 |
| Zn con | 218800 | 1.40 | 21 | 51.7 |

(a) What are the overall metal recoveries at each plant?

(b) The tails are to be blended with metal-free overburden and dumped, how much overburden is required if the dump is to have metal contents below the environmental limits (0.05% Pb, 2 g t$^{-1}$ Ag, 0.1% Zn) where no further treatment is required? This is an example of the dubious principle 'the solution to pollution is dilution'. If the plants pay for dumping on a pro rata basis which plant is subsidising the others the most?

3. A biological oxidation plant treats 100 t d$^{-1}$ of a concentrate comprising 40 wt% arsenopyrite, 55 wt% pyrite and quartz. The oxidation of the arsenopyrite is complete but only 60% of the pyrite is oxidised at the discharge. The slurry is filtered to 5 wt% moisture and the solids passed to cyanidation to recover the gold liberated from the arsenopyrite.

(a) The leach solution is maintained at pH 2 to ensure maximum biological activity, how much lime needs to be added?

(b) After filtration, the solution is to be neutralised to pH 8 for disposal, how much lime needs to be added?

(c) To ensure long-term stability of the arsenic, a non-sulphidic-iron:arsenic ratio of 4:1 in the tailings is required. How much iron needs adding to the tailings? In what form would you recommend adding it?

(d) The pyrite in the tailings will oxidise in time forming acid drainage, to circumvent this the tailings are to be mixed with limestone to neutralise any acid generated. If the acid neutralising capacity has to be double that of the acid generating capacity, how much limestone is required?

(e) When the solids are repulped to 30 wt% they have to be brought to pH10 before any cyanide is added, how much lime is needed for this?

(f) The plant decides to make its own lime from locally available limestone, what is the annual requirement of limestone?

4. A tungsten mine produces a pure scheelite-powellite concentrate which is 79.5% $WO_3$. Tungsten is produced by a process involving leaching of the ore with a 3:1 sodium carbonate: metal molar ratio at 40 wt% initial solids, decreasing the solution to pH 2 with sulphuric acid, selectively precipitating the Mo with $H_2S$, raising the pH with lime until all sulphate precipitates as gypsum, raising the pH to 13 with sodium hydroxide to precipitate the tungsten as $WO_3.H_2O$ and then hydrogen reduction to metal. The annual production of W is 250 tonnes.

(a) calculate the reagent consumption per tonne of metal produced

(b) estimate the water balance for the plant?

(c) how much Mo is produced per annum?

(d) a local factory offers to supply a 0.5 M sodium carbonate solution for free, do you accept it? Why?

5. A stream containing $10 \, g \, L^{-1}$ sulphuric acid and $1 \, g \, L^{-1}$ copper stream flows from a heap leach at $0.1 \, m^3 \, min^{-1}$ and the copper is recovered by cementation onto scrap iron. To avoid acid consumption from the stream, the copper in solution cannot be depleted below $0.05 \, g \, L^{-1}$. For every mole of copper removed below, this level a mole of protons is consumed. The residence time of solution in the heap is 7 days.

(a) how much iron is consumed per annum?

(b) if the copper is present exclusively as malachite, how much sulphuric acid needs to be added to maintain the pH?

(c) during the closure of the heap the copper is depleted to $0.001 \, g \, L^{-1}$ and lime added to bring the pH to 7 for disposal, how much lime is required?

6. The tailings from a gold plant are pumped to a tailings pond at a flow of $10 \, m^3 min^{-1}$ of pulp with an SG of 1.50 containing 45% by mass of solids, and a concentration of 5 ppm of cyanide (as CN) in the solution. A small stream also discharges into the pond at a flow of $2 \, m^3 min^{-1}$ with a concentration of 0.01 ppm cyanide. Solution is pumped back to the plant from the pond at a rate of $5 \, m^3 min^{-1}$ stream at a concentration of 0.1 ppm cyanide. Calculate the rate of loss of cyanide from the tailings pond assuming that the pond is well mixed.

# Case study

A copper-gold plant has the following monthly production figures.

| Stream | Tonnes | g t$^{-1}$ Au | % Cu |
|---|---|---|---|
| Feed | 90393 | 3.42 | 0.77 |
| Gravity conc | 6 | 694 | 51.7 |
| Gravity tails/flotation feed | 90387 | 3.37 | 0.77 |
| Flotation conc | 1300 | 186 | 47.7 |
| Flotation tails/CIL feed | 89087 | 0.73 | 0.08 |
| CIL tails | 89087 | 0.20 | 0.08 |

(a) draw the simple flowsheet for the plant that involves gravity recovery for coarse gold, flotation to recover copper and gold and CIL (carbon-in-leach) to recover residual gold from the flotation tailings.

(b) perform mass balances for gravity and flotation circuits, explain the most likely reasons for any discrepancies

(c) how much gold and copper are produced each month?

(d) what are the recoveries of gold and copper in each step?

(e) what is the total recovery of gold and copper

(f) the company shipped 49.48 kg of bullion in this month, how much silver was there in the bullion?

# 5

# Kinetics of reactions in hydrometallurgy

The reactions that occur during most hydrometallurgical processes involve heterogeneous interactions between surfaces and chemical reagents in aqueous solutions. These surfaces are most commonly solids as in leaching of minerals but can be metals in the electrowinning and refining of metals. The surfaces can also be interfaces between two non-solid phases such as the transfer of a gas into a liquid phase, the transfer of metal ions and reagents across a liquid/liquid interface in solvent extraction. The description of the kinetics of these reactions must involve both the rates of the chemical reactions at the surfaces and the transfer of reagents to and products from the surfaces.

Although, as dealt with in Chapter 3, the thermodynamics of the particular chemical reaction will determine whether it is possible for the reaction to occur, the rate (or kinetics) of the reactions will determine whether the reaction can be of practical value. This module will introduce the main concepts of those aspects of kinetics which are important in the description and understanding of hydrometallurgical processes.

It is assumed that the reader will have had an introduction to the basic concepts of chemical kinetics that will be briefly reviewed in the initial sections. The important theory associated with electrochemical reactions will be outlined in more detail as it applies to many hydrometallurgical processes. The fundamental aspects of mass transport will be reviewed and contrasted with the rates of chemical reactions. The influence of mass transport in limiting the overall kinetics and that of chemical reaction in increasing the rate mass transport will conclude this chapter.

## 5.1 Homo- and heterogeneous processes

Reactions which occur within the same phase are called homogeneous while those that take place at the interface of two

Hydrometallurgy. https://doi.org/10.1016/B978-0-323-99322-7.00004-1

or more phases are called heterogeneous. Thus, for example, the following reaction which is important in many leaching and precipitation processes

$$Fe^{2+}(aq) + O_2(aq) \rightarrow Fe^{3+}(aq) + H_2O(aq)$$

is a **homogeneous** reaction as written because all reactants are dissolved species.

On the other hand, the process

$$Fe^{2+}(aq) + O_2(g) \rightarrow Fe^{3+}(aq) + H_2O(aq)$$

$$O_2(g) \rightarrow O_2(aq)$$

is a **heterogeneous** reaction because it now includes an additional step which takes place at the gas/liquid interface.

Classify the following reactions as homo- or heterogeneous

$$Cu^{2+}(aq) + 2e = Cu(s)$$

$$Ni^{2+}(aq) + 2HL(org) = NiL_2(org) + 2H^+(aq)$$

$$Ni(NH_3)_4^{2+}(aq) + H_2(aq) = Ni(s) + 2NH_3(aq) + 2NH_4^+$$

$$Ca^{2+}(aq) + SO_4^{2-}(aq) = CaSO_4(s)$$

$$2Fe^{3+}(aq) + SO_2(g) + 2H_2O = 2Fe^{2+}(aq) + SO_4^{2-}(aq) + 4H^+(aq)$$

$$ZnS + 2Fe^{3+}(aq) = Zn^{2+}(aq) + S(s) + 2Fe^{2+}(aq)$$

## 5.2 The rate-determining step

The rate of any heterogeneous process will generally be governed by the rate of one of the steps in the overall reaction, that is, by the rate of the so-called rate-determining step. Consider the dissolution of the mineral sphalerite (ZnS) by oxidation in acidic sulphate solutions such is practiced in several modern zinc plants.

The overall reaction can be written as

$$2ZnS(s) + O_2(g) + 4H^+(aq) = 2Zn^{2+} + 2S(s) + 2H_2O(aq) \quad (5.1)$$

Although the reaction is written with gaseous oxygen as the reactant, the process does not actually require that all three phases (gas, liquid and solid) are in contact but that dissolved oxygen reacts at the mineral surface. The overall process can therefore be schematically depicted as shown in Fig. 5.1.

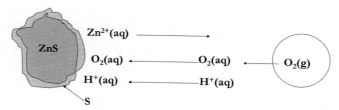

**Figure 5.1** Steps in the oxidation of a mineral.

It should be obvious that the rate of any one of the following steps could dictate the rate at which the oxidation reaction will occur.

1. Transfer of oxygen from the gaseous phase to the aqueous phase.
2. Transport of dissolved oxygen from the vicinity of a gas droplet to the surface of the mineral.
3. Diffusion of oxygen (or protons) through the film of elemental sulphur which forms as a product layer around the mineral particle.
4. The chemical reaction of oxygen (and protons) with the mineral at the mineral/sulphur interface.
5. Transport of $Zn^{2+}(aq)$ ions away from the reaction site by diffusion out through the sulphur layer.

Depending on the conditions under which the reaction is carried out, any one of the above steps could be the rate-determining step. It is obviously important to establish the rate-determining step if one is to maximise the rate of the oxidation reaction. We will return to this example in more detail at a later stage.

## 5.3 Slow chemical reactions

In many cases, the rate-determining step is the chemical reaction itself. Thus, for example, in the reaction sequence

$$1. \quad O_2(g) \rightarrow O_2(aq) \tag{5.2}$$

$$2. \quad 4Fe^{2+}(aq) + O_2(aq) + 4H^+(aq) = 4Fe^{3+}(aq) + 2H_2O(aq) \tag{5.3}$$

referred to in Section 5.1, the rate (at least at temperatures below 150–200°C) is not determined by the rate of transfer of oxygen from the gaseous phase to the aqueous phase (step 1) but by the rate of the reaction (step 2). In sulphate solutions, the rate of this step can be described by the rate equation

$$-d[Fe^{2+}]/dt = k[Fe^{2+}]^2 \cdot [O_2]/[H^+]^{0.25} \tag{5.4}$$

where [..] denotes the concentration of the species and k is a rate constant.

It should be apparent that the form of this rate equation bears no relationship to the overall stoichiometric reaction (5.3). Thus, one could be tempted to predict that the rate would vary with $[Fe^{2+}]^4$ and would also increase with $[H^+]^4$. In fact, the opposite is true for the dependence of the rate on the acidity.

Remember: The form of the rate equation for a chemical reaction cannot, except for the simplest of reactions, be predicted from the overall stoichiometric reaction equation.

For example, the overall reaction in the cyanidation process for gold can be written as

$$4Au + 8CN^- + O_2 + 2H_2O = 4Au(CN)_2^- + 4OH^- \tag{5.5}$$

but this does not imply that the rate will be eighth-order with respect to the cyanide concentration and first order with respect to the oxygen concentration. As we will see in a later section, it can be either zero or first order with respect to either depending on the conditions.

Unlike our ability to predict the thermodynamics of chemical reactions, we cannot predict the rate equation or the rate of a chemical reaction except for the simplest of reactions. These quantities have to be established by experimental studies of the rate of the process.

The rate equation for a chemical reaction is often used (with other information) to derive a mechanism for the reaction, which, in the case of the above reaction has been suggested to be made up of the following reaction steps,

$$1. \quad Fe^{2+}(aq) + H_2O \rightarrow FeOH^+(aq) \tag{5.6}$$

2. $FeOH^+(aq) + O_2(aq) \rightarrow \{FeOH^+ \bullet O_2\}(aq)$     (5.7)

3. $\{FeOH^+ \bullet O_2\}(aq) + FeOH^+ \rightarrow$ products     (5.8)

In this case, reaction (5.8) is the rate-determining step, that is is very much slower than reactions (5.6) or (5.7) which can be considered to be rapid equilibria.

## 5.3.1  Integrated forms of the rate law

If the rate law for a reaction

$$A \rightarrow Products$$

is known to be of the form

$$Rate = k[A]^n \qquad (5.9)$$

where $n$ (known as the order of the reaction with respect to reactant A is either zero, one or two, **and** the rate of the reaction depends (or can be made to depend) on the concentration of this one species, the order of the reaction can be determined graphically as shown in Table 5.1.

**Table 5.1 Characteristics of simple rate laws.**

| Order ($n$) | Rate law | Integrated rate equation | Linear plot | Slope | Units for k |
|---|---|---|---|---|---|
| 0 | rate = k | $[A]_0 - [A]_t = kt$ | [A] vs t | $-k$ | $Ms^{-1}$ |
| 1 | rate = k [A] | $\ln([A]_t/[A]_0) = -kt$ | ln [A] vs t | $-k$ | $s^{-1}$ |
| 2 | rate = k $[A]^2$ | $(1/[A]_t) - (1/[A]_0) = kt$ | (1/[A]) vs t | k | $M^{-1}s^{-1}$ |

Note that the half-life for a reaction is defined as the time required for the concentration to decrease to 50% of its initial value. Substituting $[A]_t = 0.5[A]_o$ in the above integrated equations gives the half-life for various orders in Table 5.2.

**Table 5.2 Half-life for simple reaction orders.**

| Order | Half-life ($t_{1/2}$) |
|---|---|
| 0 | $0.5[A]_o/k$ |
| 1 | $\ln2/k = 0.693/k$ |
| 2 | $1/\{[A]_o k\}$ |
| n ($\neq 1$) | $(2^{n-1} - 1)/\{k(n-1). [A]_0^{n-1}\}$ |

Thus, for a general reaction order n (not equal to 0 or 1), a plot of $t_{1/2}$ versus $[A]_0$ will give a straight line of slope n−1.

**Example:** The data for a hypothetical reaction whose rate is known to depend on the concentration of a single reactant, A (Fig. 5.2). The order of the rate law with respect to A is unknown.

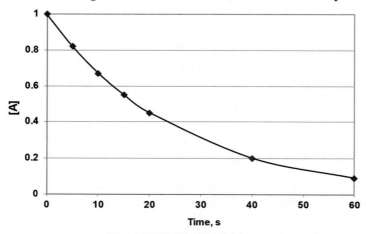

**Figure 5.2** Typical kinetic data.

This data can be used to determine the order of the reaction. We assume that the reaction is either zeroth, first or second order.

### Is the reaction zeroth order?

If the reaction is zeroth order, a plot of concentration versus time will result in a straight line. This is obviously not the case.

### Is the reaction 1st order?

If the reaction is first order, a plot of the ln of concentration versus time will result in a straight line. Since this plot in Fig. 5.3 is clearly **linear**, the reaction is probably first order.

### Is the reaction second order?

If the reaction is second order, a plot of 1/concentration versus time as shown in Fig. 5.4 will result in a straight line. It is apparently not second-order.

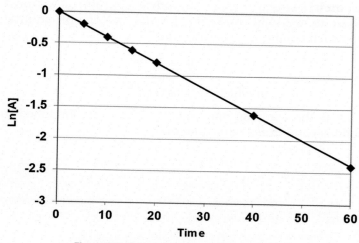

**Figure 5.3** First-order plot of data in Fig. 5.2.

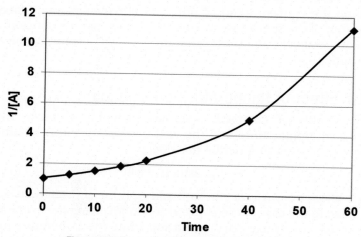

**Figure 5.4** Second-order plot of data in Fig. 5.2.

What is the half-life of the above reaction?

## 5.3.2 Simplification of rate equations

In above example of the auto-oxidation of ferrous ions (Eq. 5.1), it is possible to keep the dissolved oxygen concentration

**constant** by sparging the solution with an appropriate gas mixture and also to maintain constant pH by the use of a pH controller. Under such conditions, we can write

$$-d[Fe^{2+}]/dt = k'[Fe^{2+}]^2 \qquad (5.10)$$

in which $k' = k[O_2]/[H^+]^{0.25}$ (5.11)

that is the complex rate equation has been simplified to a pseudo-second-order rate equation.

Assuming that the pH and the dissolved oxygen concentration are maintained at 2.0 and 1 mM respectively, calculate the half-life of a solution containing 0.1 M $Fe^{2+}$ at 25°C using the above rate equation with $k = 0.015$ min$^{-2.75}$. Comment on the magnitude of $t_{1/2}$ in terms of practical use.

In some cases one can simplify a complex rate equation by making the concentration of one or more species much larger than that of the species we are monitoring so that the concentration of the species in excess is approximately constant during an experiment.

## 5.3.3 Effect of temperature on kinetics

The rates of chemical reactions generally increase with increasing temperature and the well-known Arrhenius relationship can be used to describe this effect.

$$k = A \exp(-E_a / RT) \qquad (5.12)$$

in which k is the rate constant and $E_a$ is the activation energy.

Thus, in the case of the above reaction involving the oxidation of ferrous ions by oxygen, one can substitute the Arrhenius equation for k to give the following rate equation which now includes the effect of temperature on the rate,

$$-d[Fe^{2+}]/dt = A \exp(-E_a / RT) \cdot [Fe^{2+}]^2 \cdot [O_2]/[H^+]^{0.25} \qquad (5.13)$$

If the activation energy for the above reaction is 80 kJ mol$^{-1}$, calculate the half-life of the reaction at 90°C. Assume the same parameters as the above problem but allow for the lower solubility of oxygen (0.40 of that at 25°C) at the higher temperature.

The effect of temperature on the rate of a process is often used as a diagnostic tool in determining whether the rate-determining step is the rate of a chemical reaction or mass transport. Thus, most (but not all) chemical reactions have activation energies in excess of $60-70$ kJ mol$^{-1}$ while in the case of mass transport (diffusion) the activation energy is below 40 kJ mol$^{-1}$. While useful, conclusions based solely on the activation energy should be treated with caution unless substantiated by other evidence such the effect of agitation.

## 5.4 Kinetic correlations

Unlike thermodynamics, there are no databases of kinetic properties that we can use to estimate the rates of the reactions (homo- or heterogeneous) that take place in hydrometallurgical systems. One has to resort to searching of the literature for data on the particular reactions involved under the appropriate conditions. There have been many attempts to correlate actual kinetic data to thermodynamic properties such as the overall free energy change for a reaction. However, this has not, as could be expected, yielded useful results. On the other hand, one can often correlate kinetic data for one type of reaction with that for a different reaction system. Examples of this are shown in Fig. 5.5 in which the rates of acid dissolution of a number of divalent metal oxides (at pH 1) and silicates (at pH 2) are plotted on a logarithmic scale against the rates of exchange of water molecules in the inner coordination sphere of the corresponding metals. The reactions involved are as follows:

$$M(H_2O)_6 + H_2O^* = M(H_2O)_5H_2O^* + H_2O \quad k_{sub} \qquad (5.14)$$

$$MO + 2H^+ = M^{2+} + H_2O \quad k_{diss} \qquad (5.15)$$

$$M_2SiO_4 + 4H^+ = 2M^{2+} + SiO_2 + 2H_2O \quad k_{diss} \qquad (5.16)$$

Also shown are the rates of reduction of the metal ions to metal at a cathode (Section 5.5) as given by the exchange current density ($i_o$).

The strong correlation between the reactivities for mineral dissolution and the rates of water exchange indicate that both oxide and silicate dissolution proceeds along similar pathways to water exchange in that the metal ions in the crystal lattice will have to coordinate with water molecules in the process of dissolving. This will be discussed in more detail in Chapter 6. Similarly, loss of water from an ion is required during conversion to a metal atom in the lattice during the deposition reaction. There are other

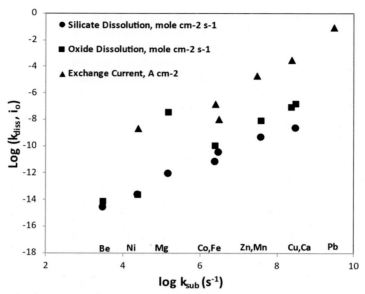

**Figure 5.5** Correlations between metal dissolution and deposition rates and the rate of exchange of water molecules for divalent metal ions. Data from Vijh and Randin, 1975; American Chemical Society (ACS), 1965; Casey and Westrich, 1992; U.S. Geological Survey (USGS), 2004.

examples in for example, solvent extraction (Chapter 10) and ion exchange (Chapter 11) in which the rates of replacement or substitution of ligands in the inner coordination sphere of metal ions are important in determining the rates of extraction of metal ions from solution.

There have been some advances in the estimation of rates of both homogeneous and heterogeneous reactions using molecular dynamics but this theory is not available to most hydrometallurgists who have to rely on actual measurement in the laboratory.

## 5.5 Electrochemical kinetics

There is a special class of chemical reactions which are important in hydrometallurgical processes. These reactions involve the transfer of electrical charge (generally electrons) across interfaces between phases. Thus, in addition to the influence of reactant concentrations on the rate of such reactions, the electrochemical potential difference between the phases (or the concentration of electrons, if you like) also has a profound effect on the rate and this is best illustrated by the following discussion of reactions at electrodes.

## 5.5.1 Reactions at electrodes

Consider the simple cell shown in Fig. 5.6 which consists of two copper electrodes in a solution or electrolyte of aqueous copper sulphate. A reference electrode has been added to the cell close to one of the copper electrodes. This enables us to measure the potential of the copper electrode with respect to the reference electrode (whose potential does not change) as a function of the current passed (which can be varied by changing the voltage V, applied by the power supply). Note that the reference electrode could be another copper electrode or any suitable reference such as an Ag/AgCl electrode. Assume that it is the former.

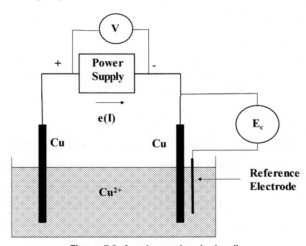

**Figure 5.6** An electrochemical cell.

For the current $I = 0$, $E_c = E_a = E_e$ (equilibrium potential) = 0.

Note that this is the potential versus the copper reference electrode in this example.

As current is allowed to flow in the direction indicated, copper will deposit on the cathode due to the reaction

$$Cu^{2+} + 2e = Cu \qquad (5.17)$$

and dissolve from the anode

$$Cu = Cu^{2+} + 2e \qquad (5.18)$$

$E_c$ will decrease and $E_a$ will increase (Why?). The overpotential ($\eta$) as defined by

$$\eta \ = \ E- E_e \tag{5.19}$$

will therefore increase at the anode and decrease at the cathode.

Let us now focus on the relationship between the current that flows and the potential of one of the electrodes relative to the copper reference electrode. However, before doing so, there are some conventions with regard to the signs of the various quantities that need to be defined.

### 5.5.1.1 Sign conventions

By the above definition and that of E, cathodic overpotentials are negative and anodic overpotentials positive.

Similarly, anodic currents are positive and cathodic currents negative.

Rate of reaction at the cathode v = No of mol of Cu deposited/unit time.

$$= \ I/nF \text{ (by Faradays law)} \tag{5.20}$$

As the process is heterogenous, the current density (i = I/A, where I is the current and A the active electrode area) is normally used.

## 5.5.2 Potential dependence of electrode kinetics

For a chemical reaction, the rate v is related to the free energy of activation ($\Delta G^{\#}$) by

$$v \ = \ kT/h \bullet \prod a \bullet \exp(-\Delta G^{\#} / RT) \tag{5.21}$$

where k and h are constants and $\Pi a$ is the product of the activities or concentrations of the reactants taking part in the rate-determining step.

For an electrode process, we have seen that $\Delta G = -nFE$ and it is reasonable, therefore, to expect the rate to be some exponential function of the potential. A more rigorous theoretical treatment enables one to derive a relationship between the rate, as reflected by the current density, and the potential of an electrochemical reaction.

Thus, for the above cell, we can draw a schematic current-potential curve for the reactions which occur as we change the potential. Thus, if we increase the voltage applied to the cell with the electron flow in the direction indicated, the left-hand electrode will become an anode and copper will be dissolved at a rate proportional to the magnitude of the current. If we reverse the polarity and increase the current in the opposite direction, the left-hand electrode will become a cathode at which copper will deposit. If we measure the potential of the electrode (relative to the copper reference electrode) at each of the currents applied, we can plot the current as a function of the potential as shown in Fig 5.7. (Remembering the sign conventions).

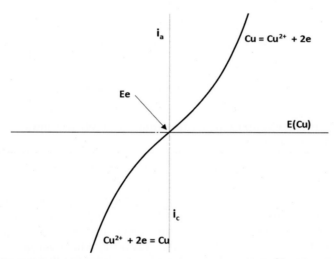

**Figure 5.7** Schematic current–potential curve for the $Cu^{2+}/Cu$ couple.

Note that the current is zero at the equilibrium potential, $E_e$.

The above current–potential relationship is, in fact, a composite of two curves as shown in Fig. 5.8.

Note that, in this case, we have changed our reference electrode to a standard hydrogen electrode and this has resulted in a shift of the equilibrium potential. The overall curve is the algebraic sum of the currents due to the anodic and cathodic half-reactions shown as dotted lines. At equilibrium ($i = 0$), the rates of both reactions are equal but opposite and the current due to each reaction at equilibrium is known as the ***exchange current density*** $(i_0)$.

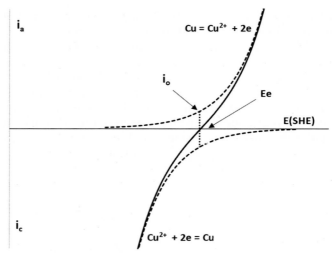

**Figure 5.8** Schematic current—potential curves for the $Cu^{2+}/Cu$ couple.

These curves can be described quantitatively as follows for the cathodic reaction,

$$i_c = -nFk_c[Cu^{2+}]\exp(-\beta_c FE/RT) \qquad (5.22)$$

where $k_c$ is a heterogeneous rate constant and $\beta_c$ (see below) a parameter known as the cathodic transfer coefficient. E is the potential with respect to any reference electrode. Note the negative exponent because the cathodic current will decrease with increasing potential and also the negative sign for the current.

Similarly, for the anodic reaction,

$$i_a = nFk_a\ \exp(\beta_a FE/RT) \qquad (5.23)$$

Note that for the above example, **[Cu] = 1 (solid)**.
At any potential, the net observed current $i = i_a + i_c$
At the equilibrium potential, $E_e$, $i = 0$ and $i_a = -i_c$

i.e. $nFk_a \cdot \exp(\beta_a FE_e/RT) = nFk_c[Cu^{2+}] \cdot \exp(-\beta_c FE_e/RT)a$
$\qquad\qquad\qquad = i_0$ (Exchange current density)

$$(5.24)$$

Thus, remembering that $\eta = E-E_e$ one can write the current/potential relationship as follows:

$$
\begin{aligned}
i &= i_0[\exp\{\beta_a F(E-E_e)/RT\} - \exp\{-\beta_c F(E-E_e)/RT\}] \\
&= i_0[\exp(\beta_a F\eta/RT) - \exp(-\beta_c F\eta/RT)]
\end{aligned}
\qquad (5.25)
$$

This is the well-known **Butler-Volmer (BV)** equation.

The transfer coefficients for a single electron transfer reaction are related by

$$\beta_c + \beta_a = 1 \text{ and for } \beta_c = \beta = 0.5(\text{generally}), \beta_a = 1 - \beta = 0.5$$

Note that, like all chemical reactions, the exchange current density is a function of the concentrations of the reactants.

For example, for the reaction,

$$Fe^{3+}(aq) + e(Pt) = Fe^{2+}(aq) \tag{5.26}$$

it has been found that

$$i_0 = i_0^0 \left[Fe^{3+}\right]^{1-\beta} \left[Fe^{2+}\right]^{\beta} \tag{5.27}$$

where $i_0^0$ is the exchange current density for all species at unit activity.

## 5.5.3 Characteristics of the Butler-Volmer equation

For $\eta > 0$, the first exponential term in the BV equation is greater than unity while the second is less than one. In other words, the net current density is positive(anodic). For $\eta < 0$ the reverse applies. At the equilibrium potential ($\eta = 0$), the rates of both processes are equal ($i_0$). A plot of the BV equation for the parameters listed is shown in Fig. 5.9, together with dotted line plots of the separate anodic and cathodic terms.

Note the current densities of 1 A m$^{-2}$ for each half-reaction at the equilibrium potential.

There are two limiting conditions under which the BV equation can be simplified.

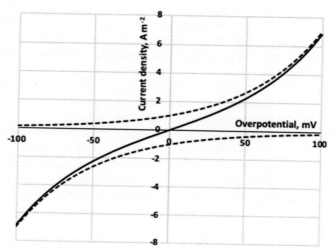

**Figure 5.9** Calculated current–potential relationship ($i_o = 1$ A/m$^2$, $\beta_a = \beta_c = 0.5$).

### 5.5.3.1 'Low-field' region

For $\eta < \pm 20$mV, there is an approximately linear region as shown in Fig. 5.9, which can be derived from $\exp(x) \approx 1 + x$ for $x \ll 1$.

Thus,

$$
\begin{aligned}
i &= i_0[1 + (1 - \beta)F\eta/RT - 1 - (-\beta F\eta/RT)] \\
&= i_0 F\eta/RT
\end{aligned}
\tag{5.28}
$$

This enables one to obtain $i_0$ using the so-called linear polarisation technique which is very common in corrosion science (Chapter 7).

### 5.5.3.2 'High-field' region

At large overpotentials, the cathodic contribution to the anodic reaction can be neglected and

$$i_a = i_0 \exp[(1 - \beta)F\eta / RT]$$

or

$$\ln i_a = \ln i_0 + (1 - \beta)F\eta/RT \tag{5.29}$$

which is the form of the empirical **Tafel** equation for an anodic process,

$$\eta_a = a + b \cdot \log_{10} i_a \quad \text{with } a = -b \cdot \log_{10} i_0$$

$$\text{i.e. } \eta_a = b \cdot \log_{10}(i_a / i_0) \tag{5.30}$$

The so-called Tafel Slope is given by $b = 2.303RT/(1-\beta)F$ and is normally given in mV/decade of current.

Note that.

**(i)** the Tafel equation is written in terms of logs to the base 10.

**(ii)** the exchange current density can be obtained by extrapolation of the Tafel line to the equilibrium potential ($\eta = 0$)

**(iii)** for a cathodic reaction, the corresponding Tafel equation is

$$\eta_c = b \cdot \log_{10}(i_0 / i_c) \tag{5.31}$$

Most electrochemical processes of importance in hydrometallurgy occur in the high-field region.

Tafel plots for the data in Fig. 5.9 are shown in Fig. 5.10.

Note the common intercept ($i = 1 \, A \, m^{-2}$) on the ordinate of the anodic and cathodic extrapolations of the linear Tafel plots. From these plots, the parameters $\beta_c$, $\beta_a$ and $i_0$ can be obtained from the slopes and intercepts of the lines.

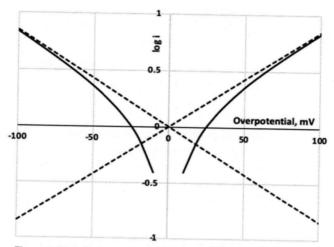

**Figure 5.10** Calculated Tafel plots using the data from Fig. 5.9.

Table 5.3 gives the electrochemical kinetic parameters for some common reactions encountered in the electrodeposition of metals. Note that the exchange current density (the electrochemical equivalent of rate constant) for a particular reaction is

**Table 5.3 Electrochemical parameters for some reactions at 25°C.**

| Reaction | Electrode | $E^0$ (V) | Solution | Tafel slope (mV) | $i_0$ (A m$^{-2}$) |
|---|---|---|---|---|---|
| $2H^+ + 2e = H_2$ | Hg | 0 | 1M H$^+$ | 120 | $5.10^{-9}$ |
|  | Zn | 0 |  | 120 | $1.10^{-6}$ |
|  | Ni | 0 |  | 120 | $1.10^{-3}$ |
|  | Pt | 0 |  | 30 | 1.0 |
| $Ni^{2+} + 2e = Ni$ | Ni | −0.26 | 1M NiSO$_4$ | 120 | $2.10^{-5}$ |
| $Zn^{2+} + 2e = Zn$ | Zn | −0.76 | 1M ZnSO$_4$ | 40 | 0.2 |
|  | Zn |  | 1M ZnCl$_2$ | 30 | 3 |
| $Cu^{2+} + 2e = Cu$ | Cu | 0.34 | 1M CuSO$_4$ | 40 | 2 |
| $Ag^+ + e = Ag$ | Ag | 0.80 | 0.1M Ag$^+$ | 60 | $1.10^4$ |
| $2H_2O = O_2 + 4H^+ + 4e$ | Pt | 1.23 | 1M H$_2$SO$_4$ | 120 | $1.10^{-4}$ |
|  | α-PbO$_2$ | 1.23 |  | 120 | $1.10^{-6}$ |
| $Fe^{3+} + e = Fe^{2+}$ | Pt | 0.77 | 0.1M Fe$^{2+,3+}$ | 120 | $4.10^3$ |

dependent on the nature of the electrode and also the composition of the solution. For reactions that involve the simple transfer of a single electron such as

$$Ag^+ + e = Ag$$

and

$$Fe^{+3} + e = Fe^{+2}$$

the rates are high with exchange current densities in excess of $1000\ A\ m^{-2}$. If two electrons are involved, the rates are somewhat lower as in the reduction of the divalent base metal ions. On the other hand, reactions that involve the breaking of bonds in addition to electron transfer, such as the oxidation of water to evolve oxygen, are slow with exchange current densities some $10^9$ times lower. The rate of reduction of metal ions from chloride solutions is generally faster than from sulphate solutions. In the case of gold (not shown) reduction to the metal from chloride solutions is several orders of magnitude faster than from cyanide solutions.

Using the data provided in Table 5.3, calculate the overpotential for the reduction of Cu ions (1 M) at a current density of 300 A m$^{-2}$. Use the 'high field' approximation.

Do the same for the evolution of oxygen on the lead anode at the same current density assuming the high field approximation.

What would be the actual potentials (versus SHE) at the anode and cathode for a solution with unit activity copper and hydrogen ions for each current density?

One of the more important electrode reactions is that for the evolution of hydrogen.

$$2H^+ + 2e = H_2$$

The rate of this reaction is strongly dependant on the surface on which the reaction takes place, and the data in Fig. 5.11 show how the exchange current density varies with metals in several rows of the Periodic Table. The platinum group elements are the most 'electroactive' while the 'soft' metals such as Zn, Cd, Hg and Pb are not good electrocatalysts for this reaction with rates some $10^{10}$ times lower than those of Pt. We shall come across such variations in several different systems.

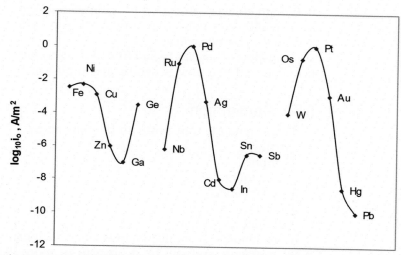

**Figure 5.11** Exchange current densities for the evolution of hydrogen on different metals. Data from Vetter, 1967.

# 5.6 Mass transport processes

As discussed in Section 5.2, the overall rate determining step may not be the chemical reaction but the rate at which material can be transported to or from the reaction site or surface. Under these conditions, we say that the rate of the reaction is mass-transport controlled. From a processing point of view, this is generally the desirable situation in that one can often increase the rate of mass transport by a variety of means, whereas a chemical reaction requires either or both of increased temperature and the use of a catalyst in order to increase the rate.

The transport of a reacting species in a moving liquid is governed by two different mechanisms. Firstly, by way of molecular diffusion as a result of concentration differences (ordinary diffusion); and secondly, by entrainment of the moving fluid transporting the solute particles (forced diffusion or convection). In turbulent regimes, most of the mass transfer is by eddy diffusion, with a small contribution from molecular diffusion. Thus, one can write

Total Mass Flux = mass transported by diffusion

$$+ \text{ mass transported by convection} \qquad 5.32$$

In turbulent regimes, it is postulated that convective transport rates are large and molecular diffusion rates can be neglected. On the other hand, for laminar flow, such as that **close to a surface** (laminar sublayer), transport is assumed to occur mainly by molecular diffusion and the contribution of eddy transport is assumed to be negligible.

Transport of species to and from a reacting surface or an interface between two phases can be accomplished by.

## 5.6.1 Convection

Hydrodynamic transport can be induced by.
Stirring
Pumping
Air Sparging
Gas Evolution
Density Gradients
Thermal Gradients

Convection is generally the most effective means of enhancing mass transport and will be dealt with in more detail at a later stage.

## 5.6.2 Diffusion

A concentration gradient acts as the driving force for diffusion from regions of high to low concentration. Diffusion is the physical process by which molecules, ions or small particles spontaneously mix, moving from regions of relatively high concentration to regions of lower concentration. It is the main contributor to mass transport close to a surface.

## 5.6.3 Migration

The contribution of migration (movement of ions under the influence of the electrical field) to the overall transport process is of importance only in processes such as mass transport to electrodes or in ion exchange materials or membranes. In theory, this can be calculated as we shall see later. In many processes, this is a rather small contribution (particularly in acidic solutions) and is therefore often neglected.

In this section, we will introduce a simple but very effective method for describing mass transport to or from a reacting surface within the aqueous phase. This surface could be a dissolving mineral particle, a gas bubble, an ion-exchange bead, an electrode or an organic droplet dispersed in an aqueous phase.

We will use the concept of a stationary liquid film adjacent to the surface. Mass transport through this film is considered to occur solely by a diffusion process (and possibly by migration in an electrical field in the case of electrochemical reactions). Beyond the film (or diffusion layer), mass transport is enhanced by means of convection involving relative movement of the surface/liquid interface by stirring or other convective processes.

Thus, Fig. 5.12 schematically depicts this model in which the concentration of a reacting species is plotted as a function of distance from the reacting surface. The diffusion layer of thickness $\delta$ is shown by the vertical line. It is assumed that the concentration is maintained by convection at its bulk value beyond the diffusion layer.

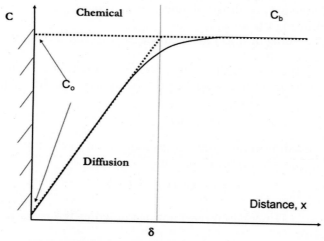

**Figure 5.12** Diffusion layer model for mass transport.

There are two limiting cases for the rate of the reaction at the surface.
**(i)** The rate is controlled by the rate of the chemical reaction. In this case, mass transport of the reacting species to the surface is very much faster than the rate of the reaction at the surface. Under these conditions, the concentration of the reactant at the surface will be essentially the same as its concentration in the bulk of the solution, that is its concentration at the surface is not perturbed by the fact that there is a chemical reaction. Under these conditions, $C_o \approx C_b$ and the concentration profile is as shown by the dotted horizontal line.
**(ii)** The rate of the reaction is controlled by mass transport of the reactant to the surface. In this case, the chemical reaction is very much faster than that of mass transport. Under these conditions, the reactant is consumed by chemical reaction as soon as it arrives at the surface. Under these conditions,

$C_0 \approx 0$ and the concentration profile is shown by the solid black line.

For a process involving diffusion, the flux (j) at the surface (or rate at which the reacting species arrives per unit area of surface) is given by Fick's First Law of Diffusion,

$$j = D(\partial C/\partial x)_{x=0} \qquad (5.33)$$

that is the rate of diffusion is proportional to the concentration gradient at the surface with D being the diffusion coefficient.

J: flux of diffusing species per unit area perpendicular to the flux direction x, [mol m$^{-2}$ s$^{-1}$]

$D$: Diffusivity or diffusion coefficient, [ m$^2$ s$^{-1}$]

x: diffusing distance, [m]

$\partial C/\partial x$: concentration gradient normal to the diffusion plane, [mol m$^{-3}$soln m$^{-1}$]

In the above case, the concentration gradient at the surface is approximated by that over the diffusion layer as shown by the dotted red line. Thus,

$$j = D(C_b - C_0)/\delta \qquad (5.34)$$

The maximum flux that can be supported under these conditions will be achieved for $C_0 = 0$,

$$j_L = C_b D/\delta = C_b \cdot k_L \qquad (5.35)$$

where $k_L = D/\delta$ is often used and is known as the ***mass transfer coefficient***.

Diffusion can be described in terms of either a diffusion coefficient or a mass transfer coefficient. The diffusion coefficient defined by this law depends on random molecular motion over small molecular distances, which means that it is a microscale mass transfer process. Diffusivity is a physical property that depends on the nature of the diffusing species and solution and is independent of flow. Most of the diffusion coefficients in liquids fall close to 10$^{-9}$ m$^2$ s$^{-1}$ while the values are considerably greater (10$^{-5}$ m$^2$ s$^{-1}$) in gases. The measurement of liquid diffusion coefficients is difficult but they can fortunately be estimated or obtained from empirical correlations.

Table 5.4 lists some typical values of the diffusion coefficients of common species in aqueous solution. Note that most metal ions have diffusivities close to 5 x 10$^{-10}$ m$^2$ s$^{-1}$ while the proton and the hydroxide ion have unusually high values for the same reason that their conductivities are high. The effect of concentration on the diffusion coefficients is not as pronounced as in the case of the conductivity and is therefore often ignored. Note

**Table 5.4 Some diffusion coefficients at infinite dilution at 25°C.**

| Species | Diffusion coefficient ($m^2\ s^{-1} \times 10^9$) |
|---|---|
| $H^+$ | 9.31 |
| $OH^-$ | 5.26 |
| $Cl^-$ | 2.03 |
| $SO_4^{2-}$ | 1.06 |
| $Na^+$ | 1.33 |
| $Cu^{2+}$, $Ni^{2+}$, $Zn^{2+}$ | 0.72 |
| $Fe^{2+}$ | 0.71 |
| $Fe^{3+}$ | 0.60 |
| $O_2$(aq) | 2.10 |
| $O_2$(g) | 23000 |
| Al in Cu | $1.3 \times 10^{-34}$ |

also the much greater rate of diffusion in the gas phase and the very low rate of diffusion in the solid state.

The thickness of the diffusion layer (and, of course the mass transfer coefficient) is strongly dependent on the effectiveness of the convective processes in the bulk of the aqueous phase. Some typical values are shown in Table 5.5.

**Table 5.5 Some approximate diffusion layer thicknesses.**

| Surface | $\delta$ (mm) | $k_L$ (m s$^{-1}$) |
|---|---|---|
| Freely suspended 100 $\mu$m particle of density 4 g cm$^{-3}$ | 0.01 | $50 \times 10^{-6}$ |
| Suspended gas bubble | 0.02 | $25 \times 10^{-6}$ |
| Plane surface with longitudinal flow (25 cm s$^{-1}$) | 0.1 | $5 \times 10^{-6}$ |
| Rotating cylinder (100 cm s$^{-1}$) | 0.04 | $12.5 \times 10^{-6}$ |
| Natural convection on vertical electrode | 0.1—0.2 | $5 - 10 \times 10^{-6}$ $5-10 \times 10^{-6}$ |

## 5.6.4 Mass transport correlations

We have seen that it is possible to estimate the rate of mass transport of a species to a surface if we know the relevant mass transfer coefficient. The coefficient will depend strongly on the geometry and flow conditions close to the surface. Thus, unlike diffusion coefficients, tables of mass transfer coefficients do not exist. A number of useful semiempirical correlations have been established which enable one to calculate the Sherwood Number, Sh (the dimensionless surface concentration gradient averaged over the mass transfer surface) for various surface geometries and operating conditions. These correlations are available in many texts on chemical engineering and use several so-called dimensionless numbers which are given in Table 5.6.

The symbols and their dimensions are:

D diffusion coefficient $(L^2 \, t^{-1})$

$k_L$ mass transfer coefficient $(L \, t^{-1})$

U fluid velocity $(L \, t^{-1})$

$\Delta\rho/\rho$ fractional density change

g acceleration due to gravity $(L \, t^{-2})$

l characteristic length (L)

$\nu$ kinematic viscosity $(L^2 \, t^{-1})$

$\mu$ viscosity $(M \, L^{-1} \, t^{-1})$

**Table 5.6 Some dimensionless groups.**

| Number(or group) | Physical significance | Form |
|---|---|---|
| Sherwood number (Sh) | Mass transfer/Diffusion | $k_L \cdot l/D$ |
| Schmidt number (Sc) | Momentum diffusion/Mass diff. | $\nu/D$ or $\mu/\rho \cdot D$ |
| Reynolds number (Re) | Inertial forces/viscous forces | $l \cdot U/\nu$ |
| Grashof number (Gr) | Buoyancy forces/viscous forces | $l^3 g \Delta\rho/(\rho \cdot \nu^2)$ |

It should be remembered that the exact definition of each of these dimensionless groups implies a specific physical system. Thus, for example, the characteristic length l in the Sherwood number will be the height from the bottom of a planar electrode for mass transfer by natural convection but will be the diameter of a spherical particle in a particulate bed or a particle of ore being dissolved.

Many of these correlations have the same general form. They involve the Sherwood number which contains the mass transfer coefficient. In the case of diffusion, the Sherwood number varies with the Schmidt number. The variation of the Sherwood number with flow is more complex because of the different physical origins of flow, that is forced convection due to pumping or other means and natural convection due to density or thermal gradients. The former involves the Reynolds number while the latter is described by the Grashof number. These and other useful correlations can be found in most chemical engineering textbooks such as those listed in the References. Some of the correlations relevant to hydrometallurgical processes are summarised below.

### 5.6.4.1 Suspended particles

This is particularly important in many hydrometallurgical operations involving mass transport to solids which are suspended in solutions or pulps such as those encountered in leaching, adsorption and precipitation or crystallisation processes.

$$Sh = 2.0 + 0.6 \cdot Gr^{1/4} \cdot Sc^{1/3} \qquad (5.36)$$

in which $Sh = k_L \cdot d/D$, $Gr = d^3 \Delta \rho g/(\rho \nu^2)$ and $Sc = \nu/D$ and $\Delta \rho$ is the density difference between particle and fluid and d is the particle diameter.

This is a particularly useful correlation and is shown in Fig. 5.13 for particles of different densities and diameters in dilute aqueous solution.

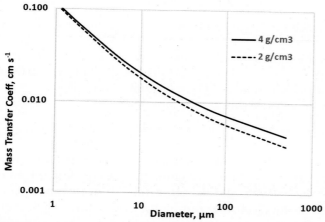

**Figure 5.13** Correlation for mass transport to suspended particles for two particle densities.

Note that the mass transfer coefficient increases as the particle size decreases and this has consequences in estimating the rate of dissolution of, for example, ore particles.

### 5.6.4.2 Natural convection on a vertical electrode

In this case the Sherwood number is given by

$$Sh = 0.67 \, (Sc \cdot Gr)^{0.25} \quad \text{for } 10^4 < Sc \cdot Gr < 10^{13} \text{ (Laminar)} \qquad (5.37)$$

$$Sh = 0.31\,(Sc\cdot Gr)^{0,28} \quad \text{for } 10^{13} < Sc\cdot Gr < 10^{15} \quad \text{(Turbulent)}$$
$$\tag{5.38}$$

$$Sh = k_L\cdot B/D \tag{5.39}$$

$$Gr = g(\rho_o - \rho_s)\,B^3/(\rho_o\,v^2) \tag{5.40}$$

in which $\rho_o$ and $\rho_s$ are bulk and surface solution densities.

$$Sc = \mu/(\rho g D) \tag{5.41}$$

in which $\mu$ = solution viscosity and $v$ = kinematic viscosity ($\mu/\rho$).

This is useful in estimating the local mass transfer characteristics (and therefore the limiting current density) at both anodes and cathodes during the electrorefining of metals.

### 5.6.4.3 Gas evolving surfaces

In this case,

$$Sh = 0.93\,Re^{0,5}\,Sc^{0,49} \tag{5.42}$$

in which $Re = V_g\cdot d\cdot\rho/A\cdot\mu$ and $V_g$ is the volumetric gas evolution rate.

d is the bubble break-off diameter (40 $\mu$m for $H_2$ evolution in alkaline solutions and $O_2$ evolution in acid solutions and 60 $\mu$m for $H_2$ evolution in acid and $O_2$ in alkaline solutions.)

In many cases, this can be simplified to

$$k_L = m\,i_g^n \tag{5.43}$$

where $m = 5{-}10.\,10^{-6}\,A^{-1}\,m^3 s^{-1}$ for $H_2$ evolution and $i_H$ in $A\,m^{-2}$

This can be used for estimating the effects of hydrogen evolution at the cathode and oxygen evolution at the anode on mass transfer at the electrodes during the electrowinning of metals.

### 5.6.4.4 Particulate beds

For a packed bed of spheres with $Re < 0{,}1$,

$$Sh = 1.09/\varepsilon\cdot(Re.\,Sc)^{1/3} \tag{5.44}$$

For a packed bed of spheres with $Re > 0{,}1$,

$$Sh = 1.44\,Re^{0,58}\,Sc^{1/3} \tag{5.45}$$

For a fluidised bed ($<25\%$ expansion),

$$Sh = (1-\varepsilon)^{1/2}\cdot Re^{1/2}\,Sc^{1/3}/\varepsilon \tag{5.46}$$

in which $Sh = k_L/U$, $Re = U\,d\,\rho/\mu$ and $Sc = \mu/(\rho.D)$ and $U =$ superficial fluid velocity (velocity in absence of particles), $\varepsilon =$ bed void age and $d =$ particle diameter.

These correlations have application for example in the description of mass transfer to ion exchange beads or activated carbon particles in columns for the recovery and separation of metal ions.

### 5.6.4.5 Rotating disk

In this case

$$Sh = 0.62\ Re^{1/2}\cdot Sc^{1/3} \tag{5.47}$$

in which $Sh = kd/D$, $Re = d^2\omega/\nu$, $Sc = \nu/D$, $d$ is the diameter of the disk and $\omega$ is the angular velocity of the disk.

This relationship is the well-known Levich equation for mass transport to a rotating disk, an important characteristic of which is that the local mass transfer coefficient is independent of the position on the disk. Rotating disks are used extensively in the laboratory study of dissolution and deposition processes in that the variation of a rate process with rotation speed can be used to distinguish between diffusion and chemical reaction rate control.

### 5.6.4.6 Suspended gas bubbles

For gas bubbles in a stirred tank,

$$Sh = 0.13\ Gr^{1/4}\cdot Sc^{1/3} \tag{5.48}$$

in which $Sh = k_L\cdot d/D$, $Gr = d^4 P_v/(\rho\nu^3)$ and $Sc = \nu/D$ and $P_v$ is the stirrer power/unit volume and $d$ is the bubble diameter.

For gas bubbles rising in an unstirred liquid,

$$Sh = 0.31\ Gr^{1/3}\cdot Sc^{1/3} \tag{5.49}$$

in which $Sh = k_L\cdot d/D$, $Gr = d^3 g\Delta\rho/(\rho\nu^2)$ and $Sc = \nu/D$ and $\Delta\rho$ is the density difference between gas and liquid and $d$ is the bubble diameter.

This is useful in describing the rate of transfer of, for example, oxygen from the gas phase into a bioleaching reactor.

Calculate the initial rate(g s$^{-1}$) of dissolution of a spherical 100 μm particle of ZnS in a 0.01 M ferric solution assuming that the rate is controlled by the mass transport of ferric ions to the surface.

### 5.6.5 Mass transport and electrochemical kinetics

The current-potential curves discussed in Section 5.3 were derived and depicted on the assumption that the rates of the reactions at the electrodes are controlled by the rates of the electrochemical processes. Thus, for example, the cathodic curve for the reduction of $Cu^{2+}$ ions shown in Fig. 5.9 shows that the current increases exponentially with decreasing potential. However, the concentration of $Cu^{2+}$ ions at the electrode surface will decrease due to reduction and at a sufficiently high overpotential, this concentration will decrease to effectively zero. Further increases in the overpotential cannot increase the rate of reduction which has now become controlled by mass transport and the curve will take on the S-shape as shown in Fig. 5.14. The maximum current that can be passed is known as the limiting current density.

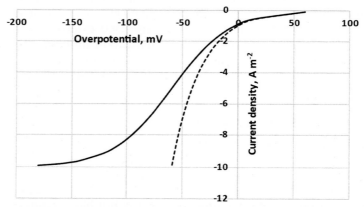

**Figure 5.14** Current–potential curve for a cathodic reaction involving mass transport.

The equation for the current-potential curve including mass transport in terms of the limiting current density $i_L$ can be shown to be the following for a cathodic reaction under high-field conditions,

$$\eta = b \cdot \{\log(i_o / i_c) - \log[(i_L - i_c)/i_L)]\} \tag{5.50}$$

and the curve will take on the shape as shown as the black line in Fig. 5.14 for $i_L = 10\,A\,m^{-2}$ and $i_o = 1\,A\,m^{-2}$.

Note that the dotted line is the cathodic branch of the Butler-Volmer equation shown in Fig. 5.9 which does not take account of mass transport. Thus, deviations from the BV equation due to mass transport limitations can occur at current densities as low as 10% of the limiting current density. We shall deal in more detail with such curves in Chapter 7.

## 5.6.6 Mass transport in leaching and adsorption

We shall deal with some specific aspects of the heterogeneous reactions involved in the leaching of mineral or metal particles in Chapter 6. Other processes such as cementation and adsorption of metal ions onto ion exchange resins or activated carbon granules are also examples of heterogenous reactions which will be dealt with in more specific detail using the principles discussed in this chapter in later chapters.

## 5.6.7 Mass transport across interfaces

The above treatment allows us to calculate the mass transport of species to or from a surface which is in a homogeneous aqueous solution. This applies to the dissolution of a solid, mass transport of ions to an electrode or to the surface of a growing crystal or precipitate. In many cases, the transport of ions or other species occurs across an interface between two phases as shown in Fig. 5.15.

In these cases, the actual transfer across the interface is generally rapid in comparison with the transport of the species to and from the interface, that is we can assume that the distribution of the transferring species between the two phases is at equilibrium *at the interface.*

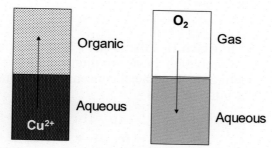

**Figure 5.15** Mass transport across liquid/liquid and gas/liquid interfaces.

We can describe mass transfer across an interface in terms of the same flux equations used above, viz

$$j = k \cdot \Delta C \qquad (5.51)$$

in which j is the flux relative to the interface, k is an overall mass transfer and $\Delta C$ is the appropriate concentration difference. The question that then arises is 'Which side of the interface is to be used for $\Delta C$?'

Consider the following simple but relevant example which has been referred to earlier in this Chapter. A gaseous reactant such as oxygen in the gas phase on the left is being transferred to the liquid phase on the right as shown in Fig. 5.16.

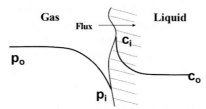

**Figure 5.16** Schematic of mass transport across a gas–liquid interface.

The flux in the gas phase is

$$j_p = k_p (p_o - p_i) \tag{5.52}$$

where $k_p$ is the gas phase mass transfer coefficient, $p_o$ is the bulk phase partial pressure of the gas and $p_i$ the pressure at the interface.

Similarly,

$$j_L = k_L (c_i - c_o) \tag{5.53}$$

In most cases, the gas and liquid phases are in equilibrium at the interface. In the case of gas/liquid equilibria, this can be expressed as

$$p_i = H \cdot c_i \tag{5.54}$$

in which H is the Henry's Law constant.

Combining the above three equations for $j_p = j_L = j$ and eliminating the interfacial concentrations, one obtains the following for interfacial flux

$$\begin{aligned}
j &= (p_o - H \cdot c_o)/(1/k_p + H/k_L) \\
&= (p_o/H - c_o)/(1/k_p \cdot H + 1/k_L)
\end{aligned} \tag{5.55}$$

This result is often compared to an electrical circuit with two resistances in series, that is j is the equivalent of the current, $(p_o - H \cdot c_o)$ is the potential difference and $1/k_p$ and $H/k_L$ are the resistances. The above relationship can be written in the form,

$$j = k(c^* - c_o) \tag{5.56}$$

where $c^* = p_0/H$ is the hypothetical liquid phase concentration of the gas that would be in equilibrium with the bulk gas and k is the overall mass transfer coefficient given by

$$1/k = 1/k_p \cdot H + 1/k_L \qquad (5.57)$$

In general, mass transfer in the gaseous phase is significantly greater than in the liquid phase. For example, the diffusion coefficient of oxygen at 25°C is 0.23 cm$^2$ s$^{-1}$ in air but $2.1 \times 10^{-5}$ cm$^2$ s$^{-1}$ in water (Table 5.4). Thus, $1/k_L \gg 1/k_p$. H for sparingly soluble gases such as oxygen which have high values for H (for oxygen H = 788 atm.L/mole at 25°C) and the above relationship can be simplified to

$$j = k_L(c^* - c_0) \qquad (5.58)$$

which can be used to calculate the rate of mass transfer from a gas to a liquid phase. The maximum achievable rate will be given when $c_0 = 0$.

Thus, for example, $c^* = 1.0 \times 10^{-3}$ M for oxygen in water at $p_0 = 1$ bar and $k_L$ is about $2 \times 10^{-3}$ cm/s, so that,

$$j_{max} = 2 \times 10^{-3} \text{ cm s}^{-1} \times 1.0 \times 10^{-3} \text{ mol L}^{-1} \times 1l/1000 \text{ cm}^3 = 2.0 \times 10^{-9} \text{ mol cm}^{-2} \text{ s}^{-1}.$$

This relatively low value could mean that mass transport becomes the rate-determining step in, for example, biological or pressure oxidation processes and emphasises the need to have a large interfacial area. Note, however, that for soluble gases such as $NH_3$, $CO_2$ and HCN, the Henry's Law H value is low ($5 \times 10^{-5}$ atm L mole$^{-1}$ at 25°C for ammonia) so that the above approximation does not always hold and in these cases, mass transport in the gas phase can be rate limiting.

Similarly, for the first example above, we can write for the flux across the liquid/liquid interface under steady state conditions,

$$j = k_a\{[Cu^{2+}]_{b,a} - [Cu^{2+}]_{s,a}\} = k_o\{[Cu^{2+}]_{s,o} - [Cu^{2+}]_{b,o}\} \qquad (5.59)$$

in which $k_a$ and $k_o$ are mass transfer coefficients in each phase, $[Cu^{2+}]_{b,a}$ is the concentration of copper ions in the bulk aqueous phase, $[Cu^{2+}]_{s,a}$ is the concentration of copper ions in the aqueous phase at the interface with corresponding quantities for the organic phase. Equilibrium at the interface further requires that $[Cu^{2+}]_{s,o} = K_d [Cu^{2+}]_{s,a}$ where $K_d$ is the distribution coefficient for the distribution of copper ions between the two phases. If we assume that $[Cu^{2+}]_{s,o} \gg [Cu^{2+}]_{b,o}$ that is we start with no copper in the organic phase, we can rearrange the above expression to give,

$$j = k_a \cdot k_o \cdot K_d \cdot [Cu^{2+}]_{b,a}/(k_a + k_oK_d) \tag{5.60}$$

that is we have an overall interfacial mass transfer coefficient

$$k_L = k_a \cdot k_o \cdot K_d/(k_a + k_oK_d) \tag{5.61}$$

If $k_oK_d \gg k_a$, then $k_L = k_a$ that is the rate of transfer is limited by mass transport in the aqueous phase.

### 5.6.8 Mass transport with chemical reaction

Diffusion rates can be significantly increased by chemical reaction. This can be seen from the data in Fig. 5.17 which shows the effective mass transfer coefficient for the absorption of ammonia gas into water at different pH values. At pH values above 5, the mass transfer coefficient is small, being typical for liquids. Below pH 4, it increases linearly with the acid concentration. This is because the reaction

$$NH_3(a) + H^+ = NH_4^+ \tag{5.62}$$

is occurring at the gas/liquid interface. Below pH 1, the coefficient again approaches a limiting value because mass transfer in the liquid has been accelerated so much that the overall mass transfer is now limited by diffusion in the gas phase.

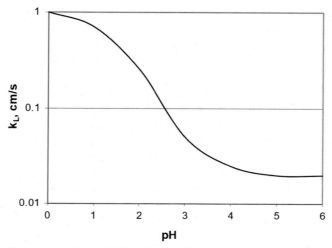

**Figure 5.17** Overall mass transfer coefficient for the absorption of ammonia gas into water at 25°C.

This effect is substantial, and it is important to allow for such phenomena in describing mass transfer effects. Consider again the simple picture of the interface in Fig. 5.18.

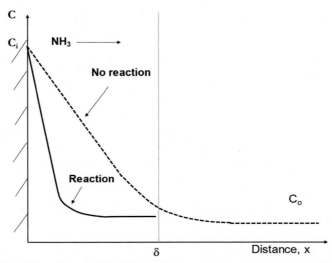

**Figure 5.18** The effect of chemical reaction on mass transfer across an interface.

In the absence of chemical reaction, the diffusion flux can be derived as before as

$$j = k_L(c^* - c_o) \tag{5.63}$$

in which the concentration at the interface $c_i = c^* = p_i/H$ and $k_L = D/\delta$ as before.

Although this was derived using the simple Nernst model of the interface, it can also be derived from Fick's First Law,

$$j = -D \cdot \partial C/\partial x \tag{5.64}$$

because $\partial C/\partial x$ is a constant value under these conditions. This is not true if the species, such as $NH_3$, is reacting within the diffusion layer. Its concentration profile will be different and can be calculated by the following procedure **for a chemical reaction of the ammonia that is first order in ammonia**.

A mass balance across a thin layer $\Delta x$ located at a point x within the layer of area A will give

Accumulation = Diffusion in − Diffusion out − Chemical
reaction

Under steady-state conditions, there is no accumulation in the layer, and one can write

$$0 = A \cdot j_{\overline{x}} - A \cdot j_x + \Delta_{\overline{x}} - k \cdot C_x \cdot A \cdot \Delta x \qquad (5.65)$$

where k is a first-order reaction rate constant.

Dividing by the volume of the thin layer, $A.\Delta x$, and rearranging,

$$(j_x - j_x + \Delta x)/\Delta x - k \cdot C_x = 0 \qquad (5.66)$$

or, for $\Delta x \rightarrow 0$,

$$\partial N/\partial x - k \cdot C_x = 0 \qquad (5.67)$$

in which $N = (j_x - j_{x+\Delta x})$ is the change in the no of moles in the thin layer per unit time.

Substituting $-j = D.\partial C/\partial x$ from Fick's law,

$$D \cdot \partial^2 C_x/\partial x^2 - k \cdot C_x = 0 \qquad (5.68)$$

Integration of this equation with the boundary conditions, $C = C_i$ at $x = 0$

and $C = C_o$ at $x = \delta$, gives the rather complex expression

$$\frac{C_x}{C_i} = \frac{\sinh[\sqrt{k/D.}\,(\delta - x)]}{\sinh[\sqrt{k/D}.\delta]} \qquad (5.69)$$

which will give the curved concentration profile shown above. The flux at any plane x is given by Fick's law as

$$j_x = -D\,\partial C_x/\partial x = \sqrt{D \cdot k} \left[ \frac{\cosh[\sqrt{k/D} \cdot (\delta - x)]}{\sinh[\sqrt{k/D} \cdot \delta]} \right] \cdot C_i \qquad (5.70)$$

At the interface, $x = 0$ and,

$$j_o = \sqrt{D \cdot k}\,\coth[\sqrt{k/D} \cdot \delta](C_i - C_o) \qquad (5.71)$$

There are two limiting conditions for this relationship.

1. For $k \rightarrow 0$, $j_o = D/\delta.(C_i - C_o) = k_L (C_i - C_o)$, which is the case for a very slow reaction which does not perturb the interfacial concentration profile.

2. For k very large, $k_L = (D \cdot k)^{1/2}$ and $j_o = (D \cdot k)^{1/2} (C_i - C_o)$, that is the rate of mass transfer is independent of agitation in the liquid phase and depends only on the rate of the chemical reaction.

## 5.6.9 Mass transport of ions

In addition to convection and diffusion, the mass transport of ions in aqueous solutions can be influenced by the presence of an electrical field and it is necessary to briefly review these additional modes of transport.

### 5.6.9.1 Conductivity

Consider two electrodes of cross-sectional area A placed a distance *l* apart in a solution of an electrolyte as shown.

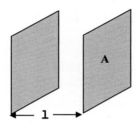

As in the case of any electrical conductor, the resistance and its reciprocal, the conductance or conductivity, is given by

Resistance : $R = \rho\, l/A$   which $\rho$ = resistivity in ohm.m   (5.72)

Conductivity : $\kappa = 1/\rho \left(\text{ohm}^{-1}\,\text{m}^{-1} \text{ or S m}^{-1} \text{ where S is the Siemen}\right)$
(5.73)

Unlike metallic conductors, the conductivity of electrolytes is concentration dependent, and therefore we define,

Molar Conductivity   $\Lambda = \kappa/c$   $\left(\text{S m}^2\,\text{mol}^{-1}\right)$   (5.74)

Generally $\Lambda$ decreases with increasing concentration and, approximately,

$$\Lambda = \Lambda^0 - Ac^{1/2} \qquad (5.75)$$

where $\Lambda^0$ is limiting molar conductivity at infinite dilution and is listed in tables.

For more concentrated solutions, one can use empirically derived relationships or compilations of such data.

Thus, in **copper** refining or electrowinning solutions, the following empirical relationship has been shown to be reasonably accurate

$$\text{Resistivity (S}^{-1}\text{cm)} = 1.1\{1.0 - 7 \times 10^{-3} (T-55) - 3 \times 10^{-3} (C_{acid} - 200) + 6.6 \times 10^{-3} C_{Cu} + 7.6 \times 10^{-3} C_{Ni} + 8.2 \times 10^{-3} C_{Fe} + 0.7 \times 10^{-3} C_{As}\}$$

with T being the temperature in °C and $C_x$ being the concentrations in g L$^{-1}$ of the various species.

For zinc electrolytes, the following equation can be used,

$$\text{Conductivity (S.cm}^{-1}) = 0.004 + 0.00115\,T + 0.0028\,[H_2SO_4].T + 0.00114\,[M].T + 0.3442$$

$$[H_2SO_4] - 0.0451\,[H_2SO_4]^2 - 0.1058\,[H_2SO_4].[M] - 0.022[M] + 0.0286\,[M]^2$$

in which [M] is the molar concentration of cations in solution as (Zn + Mg + Mn + 0.39Na + 0.23K), $[H_2SO_4]$ is the moles L−1 of sulphuric acid in solution and T is the temperature in °C.

For electrolytes (containing **nickel** and sodium sulphate) used in the electrowinning of nickel the following equation has been shown to predict the conductivity to within about 5%.

$$\text{Conductivity (S·cm}^{-1}) = -0.118 + 0.0728[Ni] - 0.00827[Ni]^2 + 0.0740[Na] + 0.0253[Ni][Na] + 0.00216T$$

in which the concentrations are in moles L−1 and T is the temperature in °C.

In the electrorefining of copper, anodes and cathodes of dimensions about 1 m × 1 m are used and the anode-cathode separation is 2 cm. Calculate the resistance between the electrodes and the voltage drop due to the electrolyte at a current density of 300 A m$^{-2}$ for an electrolyte containing 60 g L$^{-1}$ CuSO$_4$ and 175 g L$^{-1}$ H$_2$SO$_4$.

**Solution:** For this electrolyte, $\kappa = 0.581$ S cm.$^{-1}$

Interelectrode resistance

$$R = l/\kappa A$$

$$= 2 \text{ cm}/(0.581 \text{ ohm}^{-1} \text{ cm}^{-1} \times 10^4 \text{ cm}^2)$$

$$= 3.4 \times 10^{-4} \text{ ohm}.$$

$$\text{Voltage drop in electrolyte} = iR$$

$$= 300 \text{ A} \times 3.4 \times 10^{-4} \text{ ohm}$$

$$= 0.102 \text{ V}$$

For the electrolyte without added acid, the voltage drop is 2.4 V. The power saved by the use of an electrolyte containing acid is therefore significant.

Tables of some relevant electrolyte conductivities are given in the Appendices to Chapter 14.

### 5.6.9.2 Combined diffusion and migration

The total flux/unit area of an ionic species to the electrode surface is given by the sum of that due to diffusion and migration,

$$J = D(\partial C/\partial x)_{x=0} + t \cdot i/nF \qquad (5.76)$$

where $t$ is the transference number of the ion in question.

The limiting current density is therefore given by

$$i_L = nFDC_b/(1-t)\delta \qquad (5.77)$$

For example, in a 1 mol $L^{-1}$ solution of $CuSO_4$

$$t_{Cu} = 53.6/(53.6 + 80) = 0.40$$

Thus, the limiting current will be some 1.67 times that for diffusion alone.

On the other hand, a typical copper electrorefining solution also contains about 1.7 mol $L^{-1}$ $H_2SO_4$ and in this case $t_{Cu} = 0.19$ and the limiting current will be only some 1.23 times that for diffusion alone. Why is acid added if this is detrimental to mass transport of $Cu^{2+}$ ions to the cathode?

Migration is also important in the treatment of the kinetics of ion exchange processes in which an electrical field can be established inside an ion exchange bead due to the unequal rates of diffusion and migration of the exchanging ions.

## 5.7 Summary

The chemical description of these heterogenous reactions requires knowledge of the stoichiometry, thermodynamics and kinetics of the processes. This Chapter has summarised the important aspect of chemical kinetics as applied to heterogeneous systems appropriate to hydrometallurgical systems. The reader should be able to analyse kinetic data for relatively simple systems in terms of reaction orders and activation energies and suggest a rate-determining step. The important concepts of mass transport in these systems should now be appreciated and the reader should be able to choose appropriate mass transport correlations and use these to estimate the maximum rate for a given system. The application of these concepts to the kinetics of the transfer of materials between different phases should enable the reader to appreciate and utilise the concepts in later Chapters in Part II.

The theory of electrochemical reactions has been developed in terms of the anodic and cathodic half-reactions as they apply to relevant systems in hydrometallurgy. The reader should be able to analyse current-potential data and derive the important parameters such as Tafel slopes, exchange current densities and limiting current densities. These methods will be utilised in subsequent Chapters in both volumes.

## References

American Chemical Society (ACS), 1965. Mechanisms of inorganic reactions. In: Gould, R.E. (Ed.), Advances in Chemistry Series No 49 (Washington DC).

Casey, W.H., Westrich, H.R., 1992. Control of dissolution rates of orthosilicate minerals by metal-oxygen bonds. Nature 355, 157–159.

U.S. Geological Survey (USGS), 2004. A compilation of rate parameters of water-mineral interaction kinetics for application to geochemical modelling. Open File Rep. 20044–21068.

Vetter, K.J., 1967. Electrochemical Kinetics. Academic Press, London.

Vijh, A.K., Randin, J.P., 1975. Correlation between rate constants for water substitution in inner coordination spheres of metal ions and their electrochemical activities in metal deposition reactions. J. Phys. Chem. 79, 1252–1254.

## Further reading

Cussler, E.L., 2009. Diffusion. Cambridge University Press, Cambridge, UK.

Harriott, P., 1962. Mass transfer to particles: Part I. Suspended in agitated tanks. AIChE J. 8, 93–101.

Missen, R.W., Mims, C.A., Saville, B.A., 1999. Chemical Reaction Engineering and Kinetics. Wiley, New York.
Oldham, K.B., Myland, J.C., 1994. Fundamentals of Electrochemical Science. Academic Press, San Diego, California.

## Practice problems

**1.** Copper electrorefining is conducted in cells containing anodes and cathodes of dimensions 1 m × 1.5 m. The separation between anode/cathode pairs is 2 cm. The electrolyte consists of approximately 1 mol L$^{-1}$ CuSO$_4$ and 1.7 mol L$^{-1}$ H$_2$SO$_4$ ($\kappa =$ 0.581 S cm$^{-1}$).

   **(a)** Calculate the voltage drop between anode and cathode due to the resistance of the electrolyte at a current density of 300A m$^{-2}$.

   **(b)** If the electrolyte contained only 1 mol L$^{-1}$ CuSO$_4$ (K = 0.025 S cm$^{-1}$), what would be the voltage drop?

   **(c)** Calculate the annual saving in electrical power by the use of the former electrolyte for a tankhouse using a current of 100 kA (Cost of power is 15 c kWh$^{-1}$).

**2.** The following data was obtained for the anodic reaction

$$2H_2O = O_2 + 4H^+ + 4e$$

on a lead electrode in a solution containing hydrogen ions with an activity of 1 mol L$^{-1}$ at 298 K. Determine the transfer coefficient and exchange current density and comment on the number of electrons involved in the rate-determining step.

| Potential V vs Ag/AgCl Ag/AgClSCE | Current density μA cm$^{-2}$ |
| --- | --- |
| 2.0 | 0.0010 |
| 2.05 | 0.0029 |
| 2.1 | 0.0081 |
| 2.15 | 0.023 |
| 2.2 | 0.066 |
| 2.25 | 0.19 |
| 2.3 | 0.053 |
| 2.35 | 1.52 |
| 2.4 | 4.33 |
| 2.45 | 12.3 |
| 2.5 | 35.0 |

3. Magnetic alloys can be produced as thin films by codeposition of iron and nickel from sulphate solutions of iron(II) and nickel(II).

   (a) Calculate the expected composition (in mass %) of the alloy produced by controlled-potential deposition at $-1.05$ V (vs SHE) from a solution containing unit activity of both metals ions at 298K, given the following data

   | Couple | $E^\circ$ (V vs NHE) | $i_o$ (A m$^{-2}$) | Tafel slope (mV/decade) |
   |--------|------|------|------|
   | $Fe^{2+}/Fe$ | $-0.45$ | $1.10^{-4}$ | 100 |
   | $Ni^{2+}/Ni$ | $-0.25$ | $2.10^{-5}$ | 120 |

   (b) What is the total current density at the above potential?
   (c) Qualitatively describe (with reasons) the effect of current density on the alloy composition.
   (d) What other process parameter could be adjusted to increase the nickel content of the alloy?
   (e) What other cathodic reaction is likely to occur simultaneously with alloy deposition and how would the current efficiency for alloy deposition be expected to change with pH?

4. In studies of the anodic oxidation of $Fe^{2+}$ at a platinum electrode at 25°C, it was found that at a concentration of $5 \times 10^{-2}$ M ferrous ions, the limiting current density was 72 mA cm$^{-2}$. At a current density of 65 mA cm$^{-2}$, the measured overpotential was 20 mV in a still solution and 8 mV in a stirred solution. Show that this result is consistent with the existence of only concentration overpotential under these conditions.

   Assuming a diffusion coefficient for $Fe^{2+}$ in this solution of $5 \times 10^{-6}$ cm$^2$ s$^{-1}$, calculate the thickness of the diffusion layer and the mass transfer coefficient. How would you increase the limiting current density?

5. In studies of the cathodic reduction of $H^+$ on a nickel electrode at 25°C, the overvoltage was found to be a linear function of the $\log_{10}$ [current density]. For a current density of 0.15 A cm$^{-2}$, $\eta = 50$ mV while for $i = 0.55$ A cm$^{-2}$, $\eta = 100$ mV. Evaluate the Tafel parameters a and b and hence the transfer coefficient and exchange current density.

   Comment on the changes to the current/potential relationship that would occur if a platinum or a zinc electrode was used.

6. Copper dispersed in porous low-grade ore particles 0.2 cm in diameter is leached with sulphuric acid. The copper dissolves

rapidly but diffuses slowly out of the ore. The porosity can be assumed constant and the copper concentration outside the ore particles can be assumed to be low. Estimate how long it will take to remove 80% of the copper if the effective pore diffusion coefficient of cupric ions is $2.0 \times 10^{-7}$ cm$^2$ s$^{-1}$.

7.  A 500 m$^3$ tank with a height of 10 m is filled with water and is to be saturated with oxygen using a small sparger 2 cm in diameter with an oxygen flow of 1 m$^3$ min$^{-1}$. The sparger produces 0.5 cm bubbles which rise through the tank at 10 cm s$^{-1}$. How long will it take to reach 50% saturation?

8.  The extraction of copper from an aqueous solution is to be conducted using a liquid ion exchange reagent dissolved in kerosene. The equilibrium extraction data under the relevant conditions are given below.

| [Cu] in aqueous (g L$^{-1}$) | [Cu] in organic (g L$^{-1}$) |
|---|---|
| 1.0 | 0.5 |
| 3.0 | 1.5 |
| 5.0 | 2.5 |
| 10.0 | 5.0 |

The mass transfer coefficients for copper ions in the aqueous phase is $2 \times 10^{-6}$ m s$^{-1}$ and $5 \times 10^{-5}$ m s$^{-1}$ in the organic phase. A feed solution of 8.5 g L$^{-1}$ copper in an aqueous phase is contacted with an organic phase containing 1.0 g L$^{-1}$ of copper.

Calculate the overall mass transfer coefficient, the initial driving force and the mass flux of copper in each phase. Estimate the thickness of the diffusion layer in the aqueous phase if the diffusivity of copper ions is $5 \times 10^{-10}$ m$^2$ s$^{-1}$.

9.  Estimate the rate at which cyanide (as HCN) will be lost from the surface of an open gold cyanidation leach tank which has a diameter of 10 m and contains a leach pulp in which the total cyanide concentration is 200 mg NaCN L$^{-1}$ of solution at a pH value of 10. Assume that the mass transfer coefficient for transport of cyanide to the surface is $1 \times 10^{-5}$ m s$^{-1}$, Henry's law constant for HCN is $1 \times 10^{-3}$ atm M$^{-1}$ and the equilibrium constant for the protonation of CN$^-$ is $10^{9.2}$.

What would be the rate of emission of HCN if the pH was increased to 12 or decreased to 8?

If there are six tanks in series and the total cyanide concentration decreases by 10% in each tank, estimate the total cyanide emission at pH 8 and pH 10 as a fraction of the cyanide fed to the plant if the tanks each have a volume of $800 \, m^3$ and the solution flow rate is $150 \, m^3 h^{-1}$.

10. Consider a highly soluble gas such as ammonia for which the Henry's Law constant at ambient temperatures is H = $0.01$ atm $M^{-1}$. Also consider sparingly soluble gases such as $CO$, $O_2$, $H_2$, $CH_4$, $N_2$, for which H $\sim 1000$ atm $M^{-1}$.

    For adsorption of these gases into water assuming no reaction.

    What are the relative resistances of the gas and liquid films to mass transfer in each case?

    Which resistance if any controls the rate of the absorption process?

    What form of rate equation should be used in each case?

    How does the solubility of the slightly soluble gas affect its rate of absorption into water?

    In which case would chemical reaction be more helpful in increasing the rate of absorption?

    (Use values of $k_l = 1 \times 10^{-3}$ cm $s^{-1}$ and $k_g = 10$ cm $s^{-1}$).

11. Copper can be dissolved in aerated acidic aqueous solutions by the reaction

$$2Cu + O_2 + 4H^+ = 2Cu^{2+} + 2H_2O$$

    The rate of reaction is controlled by mass transport of dissolved oxygen to the surface of the copper metal. Estimate the maximum rate of dissolution (in g copper $h^{-1}$) of copper spheres suspended in a stirred reactor with an air/solution interfacial area of $1 \, m^2$. The aqueous phase mass transfer coefficient for oxygen is $1 \times 10^{-3}$ cm $s^{-1}$ and the solubility of oxygen in equilibrium with air is $2.5 \times 10^{-4}$ M.

    How would you increase the rate of dissolution?

12. Copper is electrodeposited on the walls of a cylinder through which a dilute solution containing $1.0$ g $L^{-1}$ of copper is flowed at a rate of $0.1$ L $min^{-1}$. The cylinder is a pipe with internal diameter 10 cm and length 50 cm. The mass transfer correlation for laminar flow through a short pipe is given by

$$Sh = 1.62 \, Re^{1/3} Sc^{1/3} (d/L)^{1/3} = k_L d/D$$

where $Re = dU/\nu$, $Sc = \nu/D$ and.

    d is the pipe diameter, L its length, $D = 5 \times 10^{-6}$ cm$^2$ s$^{-1}$ is the diffusion coefficient of copper ions, U is the linear flow-rate, $\nu = 0.01$ cm$^2$ s$^{-1}$ is the kinematic viscosity and $k_L$ is the mass transfer coefficient.

**(a)** Calculate the mass transfer coefficient.

(b) Assuming that the rate of deposition is controlled by mass transport of copper ions to the surface of the pipe, calculate the concentration of copper ions in the solution leaving the pipe (For a tubular reactor, $C = C_o \exp(-k\tau)$ where $\tau$ is the residence time.).

(c) Suggest two methods for increasing the rate of removal of copper in this reactor.

13. Copper can be cemented onto iron particles by the reaction

$$Cu^{2+} + Fe = Cu + Fe^{2+}$$

A dilute copper solution containing 0.50 g $L^{-1}$ of copper is flowed continuously at a rate of 1 $m^3 h^{-1}$ through a stirred tank reactor of volume 10 $m^3$ containing 10 kg of suspended spherical iron particles with a diameter of 1 mm.

The mass transfer correlation for mass transport to suspended particles is given by

$$Sh = 2.0 + 0.6\,Gr^{1/4}.Sc^{1/3} = k_L d/D$$

where $Gr = d^3 (\rho_S - \rho_L) \cdot g/(\rho_L \cdot v^2)$, $Sc = v/D$ and d is the particle diameter, g = 980 cm $s^{-2}$ is the acceleration due to gravity, $D = 5 \times 10^{-6}$ $cm^2 s^{-1}$ is the diffusion coefficient of copper ions, $\rho_S = 7.9$ g $cm^{-3}$ is the density of the solid particles and $\rho_L = 1.0$ g $cm^{-3}$ is the density of the solution, $v = 0.01$ $cm^2 s^{-1}$ is the kinematic viscosity and $k_L$ is the mass transfer coefficient.

(a) Calculate the mass transfer coefficient.

(b) Assuming that the rate of cementation is controlled by mass transport of copper ions to the surface of the iron particles, calculate the concentration of copper ions in the solution leaving the reactor.

(c) Suggest two assumptions that have been made in estimating the performance of the reactor.

14. A bubble of oxygen originally 0.1 cm in diameter is injected into excess deoxygenated stirred water. After 7 minutes, the bubble diameter is 0.054 cm. What is the mass transfer coefficient and how long will it take for 99% of the bubble to disappear?

15. A solid disk of ZnS with a diameter of 2.5 cm is spinning at 20 rpm in a large volume of an acid solution of unit activity at 25°C. Calculate the maximum rate (in g of Zn $h^{-1}$) of dissolution of the disk if $D_{Zn} = 5 \times 10^{-6}$ $cm^2$ $s^{-1}$ and $Ksp(ZnS) = 10^{-24}$ and for the reaction

$$S^{2-} + 2H^+ = H_2S(aq), \quad K = 10^{21}.$$

## Case study 1

The oxidation of ferrous ions to ferric ions in sulphate solutions by dissolved oxygen is an important reaction in the leaching of most sulphide and some oxide minerals which require oxidation of the mineral by ferric ions. The low concentration of dissolved oxygen and the very slow direct oxidation kinetics by oxygen requires a catalyst or redox mediator such as the ferric/ferrous couple in order for these reactions to proceed at an acceptable rate. The re-oxidation of ferrous to ferric by dissolved oxygen is therefore important. The rate of this reaction can be increased by increasing the temperature and the oxygen concentration (by increasing the pressure) as in the so-called pressure leach process. It can also be increased in the presence of suitable catalysts as in the bacterial oxidation or bio-leaching process. The following paper deals with the kinetics of this reaction.

Oxidation of Fe(II) by Dissolved Oxygen in Sulfate Solutions: Iwai, Majima and Awakura, Met Trans B Vol 13B, 311—318, 1982.

The rate of oxidation was measured by contacting solutions of Fe(II) in sulphuric acid with oxygen under constant pressure in a well stirred, sparged autoclave at various temperatures. Under these conditions the concentration of dissolved oxygen can be safely assumed to be constant during any one run because of the slow rate of consumption of oxygen by the reaction. Samples of the solution were taken at various times and the Fe(II) concentration analysed by titration with dichromate solution. The following tables summarise the results of these experiments.

You are required to make a study of the kinetics of this reaction using the data given as well as other appropriate data that may be required.

i.   Write a balanced chemical equation for the reaction.
ii.  Referring to the data in Table 5.7, establish the order of the reaction with respect to iron(II) and use Excel to make suitable plots from which you can derive appropriate rate constants for each run.

**Table 5.7**

| Temp | 353K | P(O$_2$) | 500 Kpa | |
| --- | --- | --- | --- | --- |
| [H$_2$SO$_4$], M | 0.1 | 0.2 | 0.5 | 1.0 |
| Time | [Fe(II)] | [Fe(II)] | [Fe(II)] | [Fe(II)] |

**Table 5.7** —*continued*

| min | M | M | M | M |
|-----|-------|-------|-------|-------|
| 0 | 0.2 | 0.2 | 0.2 | 0.2 |
| 30 | 0.135 | 0.155 | 0.17 | 0.176 |
| 60 | 0.105 | 0.127 | 0.15 | 0.16 |
| 90 | 0.087 | 0.11 | 0.131 | 0.145 |
| 120 | 0.075 | 0.095 | 0.12 | 0.133 |
| 150 | 0.064 | 0.082 | 0.11 | 0.124 |
| 180 | 0.057 | 0.074 | 0.1 | 0.116 |
| 240 | 0.045 | 0.061 | 0.087 | 0.101 |
| 300 | | 0.052 | 0.076 | 0.088 |
| 360 | | | | 0.078 |

**iii.** In a similar way, establish the reaction order with respect to oxygen by using the data in Table 5.8 and derive rate constants.

**Table 5.8**

| Temp | 353K | [H$_2$SO$_4$] | 1.0M |
|------|------|------|------|
| P(O$_2$), KPa | 700 | 500 | 300 |
| Time | [Fe(II)] | [Fe(II)] | [Fe(II)] |
| min | M | M | M |
| 0 | 0.2 | 0.2 | 0.2 |
| 30 | 0.172 | 0.176 | 0.182 |
| 60 | 0.153 | 0.16 | 0.175 |
| 90 | 0.135 | 0.145 | 0.165 |
| 120 | 0.116 | 0.133 | 0.149 |
| 150 | 0.108 | 0.124 | 0.142 |
| 180 | 0.093 | 0.116 | 0.133 |
| 240 | 0.079 | 0.101 | 0.118 |
| 300 | 0.068 | 0.088 | 0.105 |
| 360 | 0.059 | 0.078 | 0.095 |

**iv.** Repeat (iii) using the data in Table 5.9 which shows the effect of added sulphate on the rate of oxidation. Suggest possible reasons for the effect of added sulphate on the rate.

**Table 5.9**

| Temp | 353K | [H₂SO₄] | 1.0M |
|---|---|---|---|
| [Na₂SO₄], M | 0.0 | 1.0 | 2.0 |
| Time | [Fe(II)] | [Fe(II)] | [Fe(II)] |
| min | M | M | M |
| 0 | 0.2 | 0.2 | 0.2 |
| 30 | 0.176 | 0.141 | 0.089 |
| 60 | 0.16 | 0.109 | 0.069 |
| 90 | 0.145 | 0.088 | 0.055 |
| 120 | 0.133 | 0.071 | 0.042 |
| 150 | 0.124 | 0.061 | 0.035 |
| 180 | 0.116 | 0.052 | 0.029 |
| 240 | 0.101 | 0.041 | 0.023 |
| 300 | 0.088 | 0.033 | |
| 360 | 0.078 | 0.028 | |

v. Derive an overall rate equation which can be used to calculate the rate of oxidation as a function of the various concentrations and the oxygen pressure at 353K. This rate equation can then be used in a linear regression of all the data to obtain a 'best-fit' value for the rate constant.

# Case study 2

Copper(I) in ammoniacal solutions forms the stable $Cu(NH_3)_2^+$ complex ion which is rapidly oxidised by dissolved oxygen to the $Cu(NH_3)_4^{2+}$ complex. The rate of the reaction can be conveniently measured by an amperometric technique which measures a current that is proportional to the concentration of copper(I).

The following results were obtained in an experiment in which the initial concentrations are shown in the second row of Table 5.11 below. The dissolved oxygen concentration was maintained constant by sparging air into the solution.

i. Write a balanced equation for the reaction.
ii. Graphically or otherwise, demonstrate that the data in Table 5.10 is consistent with a reaction which is first order in Cu(I).
iii. Using the data in Table 5.11, derive the reaction orders with respect to ammonia and dissolved oxygen.

**Table 5.10 Results of a test with initial concentrations in second row of Table 5.11.**

| Time (s) | Current |
|----------|---------|
| 0.0 | 1.472 |
| 0.5 | 1.298 |
| 1.0 | 1.126 |
| 1.5 | 0.993 |
| 2.0 | 0.868 |
| 2.5 | 0.758 |
| 3.0 | 0.666 |
| 4.0 | 0.519 |
| 5.0 | 0.402 |
| 6.0 | 0.315 |
| 7.0 | 0.243 |
| 8.0 | 0.194 |
| 10.0 | 0.122 |
| 12.0 | 0.080 |
| 14.0 | 0.050 |
| 16.0 | 0.029 |

**Table 5.11 Pseudo-first-order rate constants.**

| [Cu(I)] (M) | [$NH_4^+$] (M) | [$NH_3$] (M) | [$O_2$] (M) | $k_1$ ($s^{-1}$) |
|-------------|---------------|-------------|------------|------------------|
| 1.00E-04 | 0.01 | 0.01 | 2.50E-04 | 0.101 |
| 1.00E-04 | 0.025 | 0.025 | 2.50E-04 | ??????? |
| 1.00E-04 | 0.05 | 0.05 | 2.50E-04 | 0.485 |
| 1.00E-04 | 0.025 | 0.025 | 2.50E-04 | 0.253 |
| 1.20E-04 | 0.025 | 0.025 | 2.50E-04 | 0.263 |
| 1.50E-04 | 0.025 | 0.025 | 2.50E-04 | 0.258 |
| 1.00E-04 | 0.01 | 0.01 | 1.00E-03 | 0.419 |

**iv.** Write the overall rate equation and calculate values for the overall rate constant for each of the experimental runs in Table 5.11.

In a pilot plant, a solution of copper(I) initially containing $1 \times 10^{-4}$ mol L$^{-1}$ of copper(I) and 0.01M ammonia/0.01M ammonium ions is oxidised by stirring the solution in an open reactor containing 1m$^3$ of solution. The internal diameter of the tank is 1 m.

Assuming that the rate of oxidation is controlled by mass transport of oxygen into the aqueous phase, calculate the rate of oxidation of copper(I) in mol l$^{-1}$ min$^{-1}$. The aqueous phase mass transfer coefficient for oxygen from the surface of the interface is $2 \times 10^{-4}$ cm s$^{-1}$ and the solubility of oxygen is $2.5 \times 10^{-4}$ mol/l.

Sketch a graph of the concentration of copper(I) in solution as a function of time under these conditions.

Compare the rate of oxidation of copper(I) under these conditions with that calculated for the rate of the chemical reaction.

# 6

# Fundamentals of leaching

The first and most important step in almost all hydrometallurgical processes is the leaching (or dissolution) of the desired metal into a suitable solution. The success of this step will determine the overall recovery of the valuable metal. For this reason, two chapters have been devoted to this important subject with the first (Chapter 6) aimed primarily at the chemistry of leaching systems and the second (Chapter 8) to the more practical engineering aspects.

Leaching involves the dissolution (preferably selective) of the valuable component(s) of an ore, concentrate, calcine or other intermediate product. The active chemical species responsible for the dissolution is known as the **lixiviant or leaching agent**.

Some common lixiviants are

**Sulphuric acid** for copper, zinc and other base metals.

**Sodium cyanide** for gold and silver.

**Hydrochloric acid/chlorine** for the platinum group metals (PGM).

**Sodium hydroxide** for bauxite (Al).

**Water** for sulphated spodumene (Li)

The terms *leaching* and *dissolution* tend to be used interchangeably although strictly speaking, the former refers to the removal (by dissolution) of a component in a mixture or compound such as an ore. The dissolved component is generally a small fraction of the total mass. This is typical of, for example the *leaching* of gold from its ores. On the other hand, the *dissolution* of zinc from a high-grade ZnO calcine involves the solution of greater than 80%–90% of the solid mass and this is strictly not a leaching but a dissolution process. We will use the terms interchangeably.

## 6.1 Types of leaching reactions

The dissolution of a solid species is a chemical reaction which can be classified into one of several types depending on the nature of the reaction. A convenient classification is given below with some examples.

Hydrometallurgy. https://doi.org/10.1016/B978-0-323-99322-7.00001-6

- **Metal Salt Dissolution**

$$NaCl(s) = Na^+(aq) + Cl^-(aq)$$

This is the simplest of dissolution processes and is used to remove residual salt from gypsum in marine salt deposits

$$NaVO_3(s) = VO_3^-(aq) + Na^+(aq)$$

This reaction is the basis for the well-known roast/leach process for vanadium ores as practiced at various plants around the world.

$$LiSO_4(s) = Li^+(aq) + SO_4^{2-}(aq)$$

This step is the final operation in the recovery of lithium into solution from ores such as those containing the silicate mineral spodumene.

- **Acid/Base Reaction**

$$ZnO(s) + H_2SO_4(aq) = ZnSO_4(aq) + H_2O$$

This process is applied throughout the world in the leaching of zinc calcines produced by the roasting of zinc sulphide concentrates.

$$Al_2O_3(s) + 2NaOH\,(aq) = 2NaAlO_2(aq) + H_2O$$

This reaction is the basis for the important Bayer process for the production of alumina

$$ZnS(s) + H_2SO_4(aq) = ZnSO_4(aq) + H_2S(g)$$

This reaction is believed to be an intermediate process in the direct pressure leaching of some zinc sulphide concentrates.

- **Ion Exchange Reaction**

$$NiS(s) + CuSO_4(aq) = CuS(s) + NiSO_4(aq)$$

This interesting reaction is the basis for some processes for the recovery of nickel and cobalt from mattes produced in the smelting of nickel/copper concentrates.

$$CaWO_4(s) + Na_2CO_3(aq) = Na_2WO_4(aq) + CaCO_3(s)$$

This reaction forms the basis for the process used to produce high purity tungsten compounds from the mineral scheelite.

- **Oxidative Dissolution**

$$UO_2(s) + 2Fe^{3+}(aq) = UO_2^{2+}(aq) + 2Fe^{2+}(aq)$$

The recovery of uranium from its ores requires the application of this reaction in practice at most uranium plants around the world.

$$CuS(s) + 2O_2(g) = CuSO_4(aq)$$

$$CuFeS_2 + 4Fe^{3+} = Cu^{2+} + 5Fe^{2+} + 2S$$

These types of reaction are becoming increasingly important in the hydrometallurgical processing of copper ores and concentrates.

$$4Au(s) + 8CN^-(aq) + O_2 + 2H_2O = 4Au(CN)_2^-(aq) + 4OH^-$$

This well-known selective reaction is the basis of the cyanidation process for gold and silver.

- **Reductive Dissolution**

$$MnO_2(s) + SO_2(aq) + 2H^+ = Mn^{2+}(aq) + H_2SO_4(aq)$$

This process has been piloted for the dissolution of manganese.

Classify the following leaching reactions as one of the above types.

1. $ZnS(s) + 2Fe^{3+}(aq) = Zn^{2+}(aq) + S(s) + 2Fe^{2+}(aq)$
2. $ZnSiO_3(s) + 2H^+ = Zn^{2+}(aq) + SiO_2(s) + H_2O$
3. $Fe_2O_3 + SO_2(aq) + 2H^+ = 2Fe^{2+} + SO_4^{2-} + H_2O$
4. $CuFeS_2(s) + Cu^{2+}(aq) = 2CuS + Fe^{2+}$
5. $CuFeS_2(s) + 2H^+(aq) = CuS + Fe^{2+} + H_2S$
6. $UO_2 + ClO_3^- + 6H^+ = UO_2^{2+} + Cl^- + 3H_2O$

# 6.2 Thermodynamics of leaching reactions

Tables of thermodynamic data or, if available, an $E_H$- pH diagram can be used to establish whether a proposed leaching reaction is thermodynamically possible and also to identify the most appropriate conditions for achieving the maximum extent of leaching (or recovery) without dissolving large amounts of impurity metal ions. In addition, one must consider the actual chemical species involved and make corrections for ionic activity coefficients if using concentrated solutions. If necessary, the Correspondence Principle can be used to estimate the leaching equilibria at elevated temperatures. On the basis of the equilibria, one can then identify appropriate conditions to maximise the extent of dissolution of the desired species and minimise dissolution of impurities. For example, one may wish to dissolve ZnO at pH values below 6 but above 3 to minimise iron dissolution.

Table 6.1 gives some examples of equilibria involved in practical leaching systems at 25°C.

One can estimate the viability in terms of the equilibrium concentrations of the desired metal ions using the above equilibrium constants. Thus, dissolution of ZnO is favoured by low pH values while dissolution of iron from hematite is unlikely even at low pH values but is favourable for FeO — a problem with iron contamination in the zinc process. The very large equilibrium constants for the oxidative leaching of gold in cyanide, heazlewoodite ($Ni_3S_2$) and uraninite mean that one does not have to take the thermodynamics into account in these cases.

The effect of temperature on the dissolution equilibrium can often be quite significant. Thus, although the equilibrium constant at 25°C for the dissolution of alumina (third row) is small, this reaction is used in the Bayer process by increasing the temperature and the activity of the hydroxyl ion.

Several examples of the application of thermodynamics in the description of leaching processes are summarised in the following sections.

**Table 6.1 Some dissolution equilibria at 25°C.**

| Reaction | Conditions | K |
|---|---|---|
| $ZnO(s) + 2H^+ = Zn^{2+} + H_2O$ | pH 3 | $1.6 \times 10^5$ |
| $ZnO(s) + H_2O + OH^- = Zn(OH)_3^-$ | pH 14 | $4.8 \times 10^{-2}$ |
| $Al_2O_3(s) + OH^- + 1.5H_2O = Al(OH)_4^-$ | pH 14 | $2.1 \times 10^{-3}$ |
| $Fe_2O_3(s) + 6H^+ = 2Fe^{3+} + 3H_2O$ | pH 1 | $6.8 \times 10^{-6}$ |
| $FeO(s) + 2H^+ = Fe^{2+} + H_2O$ | pH 1 | $3.4 \times 10^{12}$ |
| $FeTiO_3(s) + 4H^+ = Fe^{2+} + TiO^{2+} + 2H_2O$ | pH 0 | $2.5 \times 10^{-5}$ |
| $4Au(s) + O_2 + 8CN^- + 2H_2O = 4Au(CN)_2^- + 4OH^-$ | pH 11, 1mM·CN$^-$ | $5.6 \times 10^{56}$ |
| $Ni_3S_2(s) + 6Fe^{3+} = 3Ni^{2+} + 6Fe^{2+} + 2S$ | 0.1M Fe$^{3+}$ | $2.0 \times 10^{19}$ |
| $CuFeS_2(s) + 4Fe^{3+} = Cu^{2+} + 5Fe^{2+} + 2S$ | 0.1M Fe$^{3+}$ | $1.9 \times 10^{19}$ |
| $ZnS(s) + 2H^+ = Zn^{2+} + H_2S(g)$ | pH 1 | $1.1 \times 10^{-6}$ |
| $3UO_2(s) + ClO_3^- + 6H^+ = 3UO_2^{2+} + Cl^- + 3H_2O$ | pH 2, 1mM·ClO$_3^-$ | $1.2 \times 10^{90}$ |

## 6.2.1 Example 1: the cyanidation of gold and silver

Gold is a relatively unreactive metal and will not readily oxidise and dissolve in aqueous solutions unless the aurous or auric cations (+1 or +3) are stabilised by certain ligands and converted into complex ions. As the Eh-pH diagram for the simple aqueous system shows gold dissolves only at extremely oxidising potentials and/or very high pH, the oxidation products are insoluble oxides and hydroxides. Oxygen is not able to oxidise gold in the absence of stabilising ligands which is why the element is widely found as the metal in nature and is regarded as being inert under ambient conditions.

However, gold is known to be transported by solutions in the geological environment most likely complexed by chloride or sulphur compounds. From a processing perspective, ligands such as chloride have been used commercially for leaching gold and, more recently, sulphur-containing ligands like thiosulphate are being considered as having application in gold leaching.

The ability of cyanide to facilitate the dissolution of gold was known in the eighteenth century but it was not until the work of McArthur and Forrest in 1887 that the method became commercially viable. The role of cyanide is to form a gold (I) dicyanocomplex which is stable over a wide range of pH and potential. Importantly, in the presence of cyanide, gold metal is oxidised by easily available oxidants such as oxygen from air.

The dissolution of gold and silver in alkaline cyanide solutions occurs as a result of the following reactions,

$$2Au + 4CN^- + O_2 + H_2O = 2Au(CN)_2^- + H_2O_2 + 2OH^- \qquad (6.1)$$

$$2Au + 4CN^- + H_2O_2 = 2Au(CN)_2^- + 2OH^- \qquad (6.2)$$

Note that the reduction in oxygen is shown as occurring in two steps with peroxide as an intermediate. This overall reaction can be expressed as a sum of the following half-reactions.

Oxidation of Gold

$$2Au + 4CN^- = 2Au(CN)_2^- + 2e$$

$$E = -0.57 - 0.059 \ \log\{[CN^-]^2 / [Au(CN)_2^-]\} \qquad (6.3)$$

Reduction of Oxygen

$$O_2 + 2H_2O + 2e = H_2O_2 + 2OH^-$$

$$E = 0.68 - 0.059 \ pH - 0.030 \ \log\{[H_2O_2] \cdot pO_2\} \qquad (6.4)$$

$$H_2O_2 + 2e = 2OH^-$$

$$E = 1.66 - 0.059 \text{pH} - 0.030 \ \log[\text{H}_2\text{O}_2] \qquad (6.5)$$

Fig. 6.1 summarises the equilibria involved in the cyanidation reactions for gold.

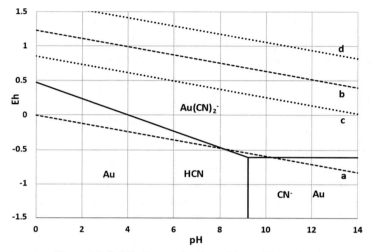

**Figure 6.1** $E_H$/pH diagram for the gold cyanide system.

The lines have been calculated using the following concentrations of soluble species which are typical of operating cyanidation plants.

Total cyanide: $4 \times 10^{-3}$ mol L$^{-1}$ or 0.02% NaCN Gold: $5 \times 10^{-5}$ mol L$^{-1}$ or 10 g t$^{-1}$ (ppm).

Note the two extra lines 'c' and 'd' to describe reactions 6.4 and 6.5.

There are no thermodynamic restrictions on the oxidation of gold in aerated cyanide solutions because the lines b, c and d all lie above the lines for the oxidation of gold to Au(CN)$_2^-$. Interestingly, hydrogen gas should be able to reduce aurocyanide to the metal at all pH values except between about 8 and 10 but this reaction is extremely slow and of no commercial significance.

An important additional acid-base equilibrium which must be considered is the protonation of cyanide

$$\text{HCN} = \text{CN}^- + \text{H}^+$$

The equilibrium can be expressed as

$$[\text{CN}^-]/\{[\text{HCN}] + [\text{CN}^-]\} = 1/(1 + K_a[\text{H}^+]) \qquad (6.6)$$

in which $K_a$ is the equilibrium constant for the deprotonation of HCN.

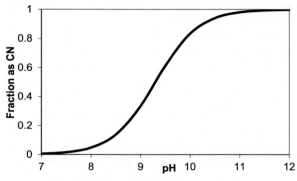

**Figure 6.2** Protonation of the cyanide ion.

Fig. 6.2 shows a plot of this equilibrium in terms of the fraction of cyanide present as the free cyanide ion as a function of pH with $CN^-$ being the predominant species at pH values to the right of the curve and HCN predominating at pH values to the left. Note that, in order to minimise loss of cyanide and exposure of operators to toxic HCN gas, the pH in the plant should be maintained at values above about 10.5.

The corresponding lines on the $E_H$-pH diagram for silver are simply shifted up by about 0.2 V in a more positive direction from those for gold.

At which pH value is the thermodynamic driving force for the dissolution of gold in cyanide the greatest? Why?

Would you operate a cyanidation plant at this pH value? Why?

Is the dissolution of gold in cyanide possible with water as the oxidant? If so, in what pH range?

## 6.2.2 Example 2: the leaching of atacamite

The mineral tenorite CuO is the simplest of the copper oxides (which includes the carbonates and silicates) but is not very common. On the other hand, the mineral atacamite $Cu_2Cl(OH)_3$ is common in many deposits especially in Chile. The chloride in the mineral results in the dissolution of chloride ions which accumulate in the circulating leach solution and can cause problems in the process. In acid, the leaching reaction is

$$Cu_2Cl(OH)_3 + 3H^+ = Cu^{2+} + Cl^- + 3H_2O \qquad (6.7)$$

The Eh/pH diagram for the copper system including atacamite for $[Cu^{2+}] = 0.05M$ and $[Cl^-] = 0.5M$ at 25°C is shown in Fig. 6.3. For simplicity, copper(I) species have been excluded. It is apparent that atacamite is more stable than $Cu(OH)_2$ except at pH values above about 10.3.

Examination of these diagrams shows that a concentration of copper of about 3 g L$^{-1}$ is in equilibrium with atacamite at a pH value of about 3.5. Thus, ignoring any kinetic effects, the pH of a heap leach process for atacamite should be below 3.5 in order to dissolve the mineral in acid solutions. This pH is lower than that required to dissolve CuO (Fig. 3.16 in Chapter 3).

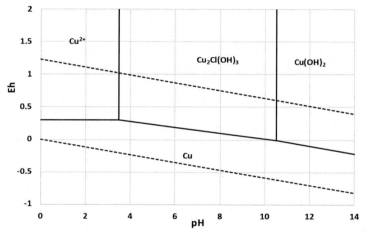

**Figure 6.3** Eh-pH diagram for the copper(II)-chloride system for $[Cu^{2+}] = 0.05M$ and $[Cl^-] = 0.5M$ at 25°C.

As a result of the stability of the copper complexes with ammonia, it is possible to dissolve both oxide and sulphide minerals in ammoniacal solutions. The diagram for the above copper system in the presence of ammonium ions is shown in Fig. 6.4.

In this case atacamite can be leached at pH values above about 7.8 to form the tetraamine complex. This diagram is also important in the leaching of copper sulphides in ammonium chloride solutions because copper will precipitate as atacamite if the pH is allowed to decrease below about 7.5. Similar diagrams can be drawn for ammonium sulphate solutions in which case the mineral brochantite, $CuSO_4 \cdot 3Cu(OH)_2$, occupies a similar region of the diagram as atacamite does in chloride solutions. In the presence of carbonate ions, malachite ($CuCO_3 \cdot Cu(OH)_2$) must be taken into account.

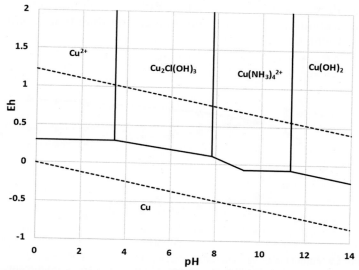

**Figure 6.4** Eh-pH diagram for copper(II) in chloride/ammonia solutions drawn for $[Cu^{2+}] = 0.05M$, $[NH_4Cl^-] = 1.0M$ at 25°C.

In these cases, it is important to recall the additional equilibria

$$CO_3^{2-} + H^+ = HCO_3^-  \quad pK = 10.3 \qquad (6.8)$$

$$HCO_3^- + H^+ = H_2CO_3(aq) \quad pK = 6.4 \qquad (6.9)$$

and

$$NH_3 + H^+ = NH_4^+ \quad pK = 9.4 \qquad (6.10)$$

which can act to buffer the pH at values close to the pK values.

## 6.2.3 Example 3: the leaching of chalcocite/ covellite

It is well known that the rate of oxidative dissolution of chalcocite ($Cu_2S$) by, for example iron(III) by the reaction

$$Cu_2S + 2Fe(III) = CuS + Cu(II) + 2Fe(II) \qquad (6.11)$$

is rapid until about 50% of the copper is dissolved after which the rate decreases substantially due to the formation of a complex mixture of sulphides within the compositional range $Cu_{1.6}S$ (geerite) to $Cu_{1.1}S$ (yarrowite). The rate of dissolution of this intermediate so-called secondary covellite is slow compared to that of chalcocite. However, the porous nature of this intermediate 'secondary covellite' is such that the rate of dissolution of this material is about an order of magnitude greater than that of natural or synthetic covellite.

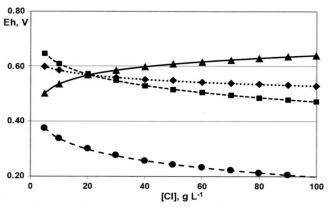

**Figure 6.5** Potentials of various redox couples as a function of chloride concentration at 25°C. Fe(III)/Fe(II) = 0.005, Cu(II)/Cu(I) = 50. ▲ Cu(II)/Cu(I), ◆ Fe(III)/Fe(II), ■ Cu(I),S/CuS, ● CuS,Cu(I)/Cu₂S.

The thermodynamics of the oxidation of covellite (written as $Cu_2S_2$) by copper(II) or iron(III) ions in chloride solutions are not as favourable as that for chalcocite, as shown by the curves in Fig. 6.5. This is similar to an Eh-pH diagram with the chloride concentration replacing the pH. The formal potentials of the couples involved

$$Cu_2S_2 = 2Cu(I) + 2S + 2e^- \qquad (6.12)$$

$$Cu(II) + e^- = Cu(I) \qquad (6.13)$$

$$Fe(III) + e^- = Fe(II) \qquad (6.14)$$

are shown as a function of the concentration of chloride ions at 25°C. Also shown for comparison is the curve for the oxidation of chalcocite to covellite,

$$2Cu_2S = Cu_2S_2 + 2Cu(I) + 2e^- \qquad (6.15)$$

The curves for the reduction of iron(III) and copper(II) have been calculated so that they intersect (i.e. in equilibrium) at a potential of 580 mV at a chloride concentration of 20 g $L^{-1}$. At equilibrium this requires that the ratios of the concentrations are Fe(III)/Fe(II) = 0.005 and Cu(II)/Cu(I) = 50.

As in Eh-pH diagrams, the requirement for spontaneous oxidation of a mineral is that the potential for the oxidant couple be greater than that for the oxidation of the mineral. Thus, at

580 mV, this is satisfied at all chloride concentrations for chalco-cite (lowest curve) but only at chloride concentrations greater than 20 g $L^{-1}$ for covellite. At a potential of 500 mV, covellite cannot be oxidised at any chloride concentration but there are no such restrictions on the oxidation of chalcocite to covellite. It is apparent, therefore, that the oxidation of covellite becomes more favourable with increasing chloride concentration for both copper(II) and iron(III) as the oxidants. Thus, in solutions of low chloride concentration, complete oxidation of chalcocite to solu-ble copper will only be possible at relatively high potentials such as above 580 mV for a solution of 20 g $L^{-1}$ chloride. At lower po-tentials under these conditions, any secondary covellite formed would be stable and the reaction would cease after approximately 50% of the copper from chalcocite is dissolved. Thus, using the data for the metastable species, the standard reduction potential for the reaction

$$Cu_{2-x}S + xCu^{2+} + 2xe = Cu_2S \qquad (6.16)$$

is 0.515 V for x = 0.33 (geerite) and 0.527 V for x = 0.62 (spion-kopite) compared to the value of 0.540 V for x = 1 (covellite).

Similarly the potential for the reaction

$$(2 - x)Cu^{2+} + S + 2(2 - x)e = Cu_{2-x}S \qquad (6.17)$$

is 0.571 V for x = 0.33 (geerite) and 0.578 V for x = 0.62 (spion-kopite) compared to the value of 0.583 V for x = 1 (covellite). Thus, these intermediate species have stabilities that are similar to that for covellite and therefore only that for covellite has been included in Fig. 6.5.

Similar considerations apply to the dissolution of digenite ($Cu_{1.8}S$) after about 45% dissolution of the copper.

We shall deal in more detail with other leaching equilibria in later sections.

## 6.3 Kinetics of leaching reactions

### 6.3.1 Rate-determining step

The dissolution of a solid is a **heterogenous** process and there-fore the rate can be expected to be governed by one or more of the following steps.

- Mass transport of the lixiviant from the bulk solution to the particle surface.
- Mass transport of products away from the particle surface.

- Mass transport within pores and cracks to the desired mineral surface.
- Mass transport within a reaction product layer around the particle.
- Chemical reaction at the mineral surface.

These steps can be illustrated as shown in Fig. 6.6 for the oxidative dissolution reaction

$$ZnS(s) + 2Fe^{3+}(aq) = Zn^{2+}(aq) + S(s) \qquad (6.18)$$

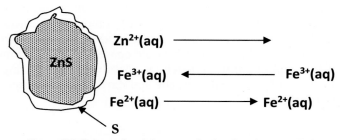

**Figure 6.6** Schematic of the steps in the dissolution of ZnS.

## 6.3.2 Behaviour of particles during leaching

Consider a spherical particle of radius R consisting of a non-porous solid B which dissolves by reaction with a lixiviant A having a bulk concentration $C_A$ which is constant with time.

$$A(aq) + bB(s) = \text{Soluble products}$$

Reaction can only occur on the surface of the particle and it will shrink with time as shown in Fig. 6.7.

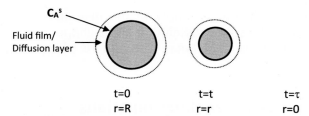

**Figure 6.7** Kinetics of the dissolution of a spherical particle.

The diffusional flux of A to the particle at time t is given by

$$F = k_f(C_A - C_A^S)4\pi r^2 \qquad (6.19)$$

where F is the flux of A to the surface of the particle.

$k_f$ is the aqueous phase mass transfer coefficient.
$C_A$ is the bulk aqueous phase concentration of A.
$C_A^S$ is the concentration of A at the particle surface.
r is the radius of the particle at time t.

Assuming that the reaction is first order in A the rate of the reaction can be written as

$$R_A = kC_A^S \cdot 4\pi r^2 \qquad (6.20)$$

in which $R_A$ is the rate in terms of moles A consumed per unit time per particle and k is a surface chemical rate constant with the same dimensions as a mass transfer coefficient.

In terms of the disappearance of the solid B, the rate can be written as

$$R_B = -dN_B/dt = bR_A = bkC_A^S \cdot 4\pi r^2 \qquad (6.21)$$

where $N_B$ is the no of moles of B in the particle at any time t.

Now

$$N_B = 4/3 \cdot \pi r^3 \cdot \rho_B \text{ and } dN_B/dr = 4\pi r^2 \rho_B \qquad (6.22)$$

where $\rho_B$ is the molar density of B.

Eq. (6.22) can be combined with $dN_B/dt = dN_B/dr \cdot dr/dt$ to give

$$R_B = 4\pi r^2 \ \rho_B \ dr/dt \qquad (6.23)$$

The mass-fraction of B dissolved (or conversion), X is proportional to the volume consumed,

$$1 - X = (r/R)^3 \text{ and } X = 1 - (r/R)^3 \qquad (6.24)$$

To solve Eq. (6.23) and obtain r and X as a function of time, consider two limiting cases.

### 6.3.2.1 Chemical reaction is rate limiting

In this case $C_A^S = C_A$ and equating (6.21) and (6.23) we get,

$$bkC_A 4\pi r^2 \ dt = 4\pi r^2 \rho_B \ dr \qquad (6.25)$$

which can be integrated with initial conditions r = R at t = 0, to give

$$r = R - bC_A/\rho_B \cdot kt \qquad (6.26)$$

The time $\tau$ to completely consume the particle (r = 0) is given by

$$\tau = \rho_B R/(bC_A k) \qquad (6.27)$$

which can be substituted in (6.26) and rearranged to give,

$$t/\tau = 1 - r/R = 1 - (1 - X)^{1/3}$$

or

$$X = 1 - (1 - t/\tau)^3 \tag{6.28}$$

### 6.3.2.2 Aqueous phase diffusion is rate limiting

In this case, $C_A^S = 0$ and combining Eqs. (6.21) and (6.23) (with allowance for the reaction stoichiometry)

$$bk_f \ C_A \ dt = \rho_B \ dr \tag{6.29}$$

Integration and re-arrangement as before gives

$$t/\tau = 1 - (r/R) = 1 - (1 - X)^{1/3} \tag{6.30}$$

with $\tau = \rho_B \ R/(bk_f C_A)$, that is the same as the expression for chemical reaction control.

Note, that $k_f$ is a function of particle size and therefore of time. Several experimental correlations have shown that, in general, $k_f$ varies with $r^n$ with n close to $-0.5$ so that we can write,

$$k_f = \alpha \cdot r^n \tag{6.31}$$

where $\alpha$ is a constant.

Substitution in Eq. (6.29) and integration and rearrangement as before gives

$$t/\tau = 1 - (r/R)^{1-n} = 1 - (1 - X)^{(1-n)/3} \tag{6.32}$$

where

$$\tau = \rho_B R^{1-n}/(b\alpha C_A(1 - n))$$

For the common case $n = -0.5$,

$$t/\tau = 1 - (1 - X)^{0.5} \tag{6.33}$$

The relationship of X versus t is thus different for these two limiting cases and is often used as a diagnostic tool in deciding which process is rate limiting.

Activation energies, obtained from the relationship of k or $k_f$ with temperature can substantiate such conclusions as they are generally less than about 20–25 kJ/mole for mass transfer control but in excess of 40 kJ/mole for chemical reaction rate control.

### 6.3.2.3 Effect of product layers

Many leaching reactions produce adherent product layers on the surface of the particles such as

$$ZnS + 2Fe^{3+} = Zn^{2+} + 2Fe^{2+} + S(s) \qquad (6.34)$$

In this case, diffusion of the lixiviant through the sulphur product layer is often the rate-limiting step. If we assume that the formation of the layer does not change the particle size and that chemical reaction is not rate-limiting, we can extend the above treatment as follows.

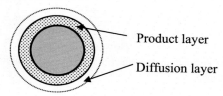

Product layer

Diffusion layer

Assuming that diffusion through the product layer is rapid compared to the movement of the front and that the concentration gradient across the product layer is linear, the diffusion flux across the layer can be written as

$$F = 4\pi \cdot rR \cdot D_A (C_A - C_A^S)/(R - r) \qquad (6.35)$$

in which $D_A$ is the diffusivity of A and the product rR is used to derive an 'average area' across which the diffusion occurs.

For rate-limiting diffusion across the product layer, $C_A^S = 0$ and Eq. (6.35) can be combined with Eq. (6.23) as before to give,

$$\{b \ 4\pi D_A \ rR \ C_A /(R - r)\} \ dt = 4\pi r^2 \rho_B dr \qquad (6.36)$$

which can be integrated to give

$$t/\tau = 1 - 3(r/R)^2 + 2(r/R)^3$$

or

$$t/\tau = 1 - 3(1 - X)^{2/3} + 2(1 - X) \qquad (6.37)$$

where

$$\tau = \rho_B R^2/(6bC_A D_A)$$

Note the different functional form and also the dependence of t on $R^2$ rather than R as found for those cases in which no product layer forms.

The shapes of the various rate equations are shown in Fig. 6.8 from which it can be seen that it could often be expected to be

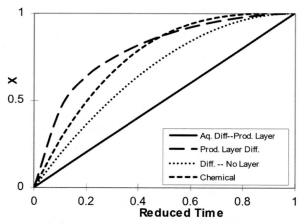

**Figure 6.8** Extent of dissolution of a spherical particle with time for various rate-determining steps.

difficult (except for the first case) to assign a particular rate-determining step on the basis of functional form only given the experimental variability in dealing with particulate systems and given the additional factors to be described in the next section.

Show that for rate-limiting aqueous film diffusion to a particle with a product layer

$$X = t/\tau \quad \text{and} \quad \tau = \rho_B R/(3bC_A k_f)$$

For rate-limiting chemical reaction, the equation is identical to that derived under (i) above. Why?

The variation of the time for complete dissolution with the particle size can also be used to distinguish between some of the above cases as shown in Table 6.2.

The assumptions made in deriving the above rate equations must always be borne in mind. In particular, the fact that a uniform particle size is used and that the concentration of the lixiviant should be maintained at a constant value are often overlooked with consequent incorrect conclusions.

The most reliable (but not conclusive) method of distinguishing between mass transport and chemical reaction as the rate determining step is the activation energy ($E_a$) obtained from measurement of the rate at different temperatures. Values of $E_a$

**Table 6.2 Effect of particle size on the time for complete dissolution.**

| Model | Rate-controlling process | Variation of $\tau$ |
|---|---|---|
| Shrinking particle without product layer | Chemical reaction | R |
| | Aqueous phase diffusion | R |
| Shrinking particle with product layer | Chemical reaction | R |
| | Aqueous phase diffusion | R |
| | Product layer diffusion | $R^2$ |

greater than about 25 kJ/mole indicate a rate determining step involving a slow chemical reaction.

## 6.3.3 Dissolution of groups of particles

In practice, we deal with groups of particles which are almost invariably not mono-sized and, as Table 6.2 shows, the dissolution rate depends strongly on the particle size and the rate determining step can even depend on the size range.

A group of particles may consist either of particles of the **same size** or of **particles distributed in size** classes. In a batch leach, each size class will dissolve at a different rate and in continuous leaching reactors, each size class may have its **own residence time distribution.** In addition, in some cases, the **chemical reactivity** may depend on the initial size of a particle and, as a result of **morphological changes** induced by the leaching reaction, the particle shape can vary with extent of reaction. A more serious problem is that real particles are seldom pure enough to be treated using these simple models. For example, a relatively 'pure' material such as a zinc calcine will contain between 75 and 90% ZnO, a base-metal matte granule 40% Ni and 30% Cu and, at the other end of the scale, a gold ore only 5 g/t of gold which occurs in various forms from fully liberated gold particles of various shapes and sizes to partially exposed inclusions in other mineral particles. Similar considerations apply to copper ores which not only have a wide mineral particle size distribution but invariably contains many different copper minerals with widely varying reactivity. This is illustrated by the QEMSCAN images of particles of a copper concentrate with chalcopyrite shown as green particles in Fig. 6.9. Note the wide size distribution (the

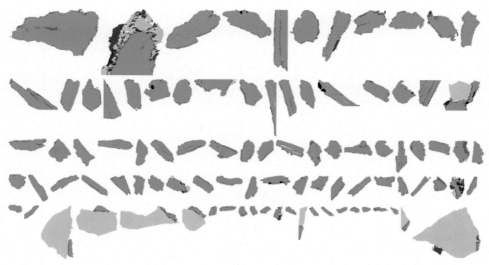

**Figure 6.9** Partial mineral map of a copper concentrate with chalcopyrite (green) particles.

length of the largest particle is about 120 μm) and also the variable particle shape.

These complications can, in theory, be included in the description of the dissolution kinetics of groups of particles but the effort is generally not justified given the many assumptions which have to be made. As a result it is not surprising that practicing hydrometallurgists have resorted to largely empirical methods to describe their leaching systems and many published so-called fundamental models should be viewed with scepticism given the adjustable parameters (which cannot be independently derived) used in the models.

### 6.3.4 Some rate equations for the leaching of gold from ores

A number of empirical and semiempirical rate equations have been used to describe the rate of dissolution of gold from various ores in batch experiments. Some of these, in increasing order of complexity are:

$$1. \quad -d[Au]/dt = k_1\{[Au] - [Au]_f\} \qquad (6.38)$$

where $[Au]_f$ represents the gold concentration in the solid phase after infinite leaching time and $k_1$ is a rate constant which is a function of the cyanide and oxygen concentrations. It has been suggested that $k_1 = a + b[CN^-] + c[O_2]$ is appropriate. An alternative which has been generally found to more describe batch leaching data is

$$-d[Au]/dt = k_2\{[Au] - [Au]_f\}^2 \qquad (6.39)$$

where $[Au]_f$ and $k_2$ have the same significance as above.

2.    $-d[Au]/dt = k_3[CN^-]^{0.8}[O_2]^{0.7}\{[Au] - [Au]_f\}^{1.5} \qquad (6.40)$

where in this case, the effects of the cyanide and oxygen concentrations are included in the rate equation.

> How does this rate equation compare with that which we could expect from our understanding of the rate-determining step(s) in the cyanidation of gold as outlined in the next section?

3.    $-d[Au]/dt = \{[Au] - [Au]_f\}\exp\{a([Au] - [Au]_f) - b\} \qquad (6.41)$

where a and b are adjustable parameters which are functions of the cyanide and oxygen concentrations.

It should be noted that (1) and (2) can be used in their integrated forms and are therefore suitable for application to batch leaching and are relatively easily used to derive the two kinetic parameters from batch leaching experiments. All can be used for continuous leaching provided the relevant parameters can be derived from batch leaching data.

## 6.4 Mechanisms of dissolution processes

The transfer of ions or atoms from a solid lattice in which they are coordinated to other ions or atoms into an aqueous solution where they are coordinated to water molecules or other complexing agents is a complex process and the mechanistic details of such reactions are still not completely resolved. It is important to recognise that the stoichiometry of a reaction does not reveal either the rate-determining step or the mechanism of the reaction.

For example, the overall reaction in the cyanidation process for gold can be written as

$$4Au + 8CN^- + O_2 + 2H_2O = 4Au(CN)_2^- + 4OH^- \qquad (6.42)$$

but this does not imply that the rate will be eighth-order with respect to the cyanide concentration and first order with respect to the oxygen concentration. As we will see, it can be either zero or first order with respect to both depending on the conditions.

It is not possible *a priori* or by reference to other similar systems to predict what the rate-determining step will be and therefore possible mechanisms for the reactions.

### 6.4.1 Oxidative leaching processes

An oxidative (or reductive) dissolution reaction can be treated as two coupled electrochemical reactions which, under freely dissolving conditions, occur at equal rates.

For example, the process

$$ZnS(s) + 2Fe^{3+}(aq) = Zn^{2+}(aq) + S + 2Fe^{2+}(aq) \qquad (6.43)$$

can be written (and in reality occurs) as the following coupled anodic and cathodic reactions

$$\text{Anodic}: \quad ZnS(s) = Zn^{2+}(aq) + S + 2e \qquad (6.44)$$

$$\text{Cathodic}: \quad 2Fe^{3+}(aq) + 2e = 2Fe^{2+}(aq) \qquad (6.45)$$

Fig. 6.10 shows schematic $i/E$ curves for the anodic dissolution of a metal sulphide MS and the cathodic reduction of $Fe^{3+}$ on the MS surface. In this, so-called Type I system, the potential at which $i_a = i_c$, is called the mixed-potential ($E_m$) and the value of $i_a$ (or $i_c$) at $E = E_m$ is the rate of dissolution($i_d$) of MS.

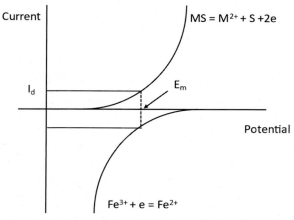

**Figure 6.10** Type I mixed-potential model for the oxidative dissolution of a metal sulphide.

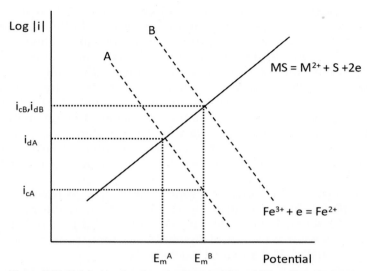

**Figure 6.11** Tafel plots for the mixed potential model in a Type I system.

If it is assumed that both anodic and cathodic reactions are occurring in the high-field (Tafel) region then the i/E curves will give linear Tafel(log) plots as shown in Fig. 6.11. The point of intersection of the anodic and cathodic lines defines $E_m$ and $i_d$. The two cathodic lines are drawn for two different concentrations of $Fe^{3+}$. If, as is expected, the cathodic reaction is first order in $[Fe^{3+}]$, then at any potential, say $E_m^A$ as shown,

$$i_c \alpha [Fe^{3+}] \text{ or } i_{cA}/i_{cB} = [Fe^{3+}]_A/[Fe^{3+}]_B \qquad (6.46)$$

Note, however, that $i_{dA}/i_{dB} \neq^1 [Fe^{3+}]_A/[Fe^{3+}]_B$ and assuming that the anodic and cathodic Tafel slopes are equal, a simple geometrical argument can be used to show that

$$
\begin{aligned}
\log(i_{dA}/i_{dB}) &= 0.5 \; \log(i_{cA}/i_{cB}) \\
&= 0.5 \; \log([Fe^{3+}]_A/[Fe^{3+}]_B) \qquad (6.47)
\end{aligned}
$$

that is the rate of dissolution is proportional to the square root of the $Fe^{3+}$ concentration.

This relationship can, of course, be derived by coupling the appropriate high-field approx. of the Butler-Volmer equation for each reaction by equating the absolute values of the currents to derive $E_m$ and thereafter $i_d$.

The above example is typical of cases where the high-field approximation is appropriate to both processes in the region of the mixed potential or a Type I system. This approximation is generally true for the anodic (or cathodic) characteristics of minerals (sulphides or oxides) which invariably exhibit highly irreversible electrochemical behaviour. However, the electro-chemical characteristics of the oxidising (or reducing) reagent are sometimes considerably more reversible even on mineral surfaces and one cannot always make the above assumption which implies that the reverse reaction (oxidation of $Fe^{2+}$ in the above example) can be neglected in the region of the mixed potential.

As shown in Fig. 6.12 for a Type III system, $E_m$ is fixed by the ratio $[Fe^{3+}]/[Fe^{2+}]$ because the contribution from the mineral anodic reaction is negligible in determining $E_m$. In this case it can be shown that the rate of dissolution is proportional to the ratio $([Fe^{3+}]/[Fe^{2+}])^{0,5}$ for equal Tafel slopes.

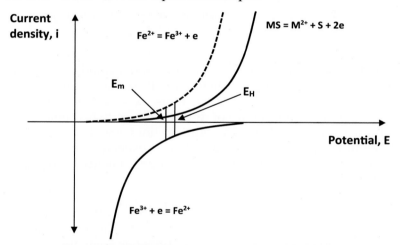

**Figure 6.12** Mixed potential model for a Type III system.

A more complicated system would be the general case where the full Butler-Volmer equation would have to be used for both anodic and cathodic reactions (Type II).

In many cases, the rate of dissolution is determined by the rate of mass-transport of the oxidising species (such as dissolved oxygen in the case of gold cyanidation) to the mineral or metal surface. This case (Type IV) is shown in Fig. 6.13.

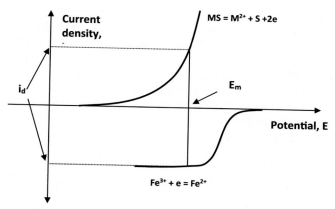

**Figure 6.13** Mixed potential model for a Type IV system.

Table 6.3 gives some examples of published reaction rate orders for various systems.

What is the rate of dissolution of a mineral MS in mol m$^{-2}$ h$^{-1}$ if the current density at the mixed potential is 10 µA cm$^{-2}$ ?

**Table 6.3 Some oxidative and reductive leaching systems.**

| Reaction | Medium | Reactants (Order) | Type |
|---|---|---|---|
| *Oxidative* | | | |
| $UO_2$ + Fe(III) | Sulphate | Fe(III) (0.5) | I |
| FeS + Fe(III) | Sulphate | Fe(III) (0.5), Fe(II) (−0.5) | III |
| $FeS_2$ + $O_2$ (acid) | Sulphate | $O_2$ (@0.5) | I |
| ZnS + Fe(III) | Sulphate | Fe(III) (0.45), Fe(II) (−0.5) | III |
| $CuFeS_2$ + Cu(II) | Chloride | Cu(II) (0.54), Cu(I) (−0.5) | III |
| PbS + $O_2$ | Alkali | $O_2$ (0.5) | I |
| Au + CN$^-$ + $O_2$ | Cyanide | CN$^-$ (0 or 1), $O_2$ (1 or 0) | IV |
| *Reductive* | | | |
| $MnO_2$ + $SO_2$ | Sulphate | $SO_2$ (0.5) | I |
| FeOOH + $SO_2$ | Sulphate | $SO_2$ (0.5−0.8) | I |
| $ZnFe_2O_4$ + Fe(II) | Sulphate | Fe(II) (0.6), Fe(III) (−0.5) | III |

### 6.4.1.1 Difference between solution and mineral potentials

As indicated above, the mixed potential of the mineral is not the same as the potential of the solution as measured by an inert electrode such as Pt.

Assuming that the oxidation of the mineral is irreversible, that is the cathodic reduction of the mineral can be ignored at the potentials encountered during oxidation, one can treat the above situation in the following way for iron(III) as the oxidant.

At the mixed-potential $(E_m)$,

$$i_a + i_m = -i_c \tag{6.48}$$

and writing each of these currents in terms of the Butler-Volmer equation,

$$i_m = k_m \exp(b_m \cdot E_m) \tag{6.49}$$

$$i_a = k_a[Fe(II)] \exp(b_a \cdot E_m) \tag{6.50}$$

$$-i_c = k_c[Fe(III)] \exp(-b_c \cdot E_m) \tag{6.51}$$

where $k_m$, $k_a$ and $k_c$ are electrochemical rate constants and $b_m$, $b_a$ and $b_c$ are the Tafel parameters $E_m$ is the potential with respect to some reference electrode and the values of k will depend on the choice of this reference potential.

These expressions can be substituted in Eq. (6.48) and the resulting equation solved for $E_m$. An analytical solution is only possible for $b_m = b_a = b_c = b$. This is not unreasonable given that the transfer coefficients are often close to 0.5.

For this case, it can be shown that,

$$E_m = \frac{1}{2b} \ln \frac{k_c}{k_a} + \frac{1}{2b} \ln \frac{[Fe(III)]}{k_m/k_a + [Fe(II)]} \tag{6.52}$$

The equilibrium potential $(E_H)$ for the iron(III)/iron(II) couple can be derived in the same terms by setting $i_a = -i_c$ for $E = E_H$ and solving for $E_H$ as above,

$$E_H = \frac{1}{2b} \ln \frac{k_c}{k_a} + \frac{1}{2b} \ln \frac{[Fe(III)]}{[Fe(II)]} \tag{6.53}$$

which, as expected, has the same form as the Nernst equation with

$$E_o = 1/2b \cdot \ln(k_c / k_a). \tag{6.54}$$

Thus, one can write

$$E_H - Em = \frac{1}{2b} \ln \left\{ \frac{k_m}{k_a[Fe(II)]} + 1 \right\} \tag{6.55}$$

which shows that, as expected, $E_H$ will be greater than $E_m$ and that, for $k_a[Fe(II)] \gg k_m$, $E_H \cong E_m$ and therefore defines the conditions under which the assumption discussed above will be valid. On the other hand, for very low concentrations of Fe(II) and/or $k_m \gg k_a$, the above relationship simplifies to

$$E_H - E_m = 1/2b \ \ln(1 / [Fe(II)]) + 1/2b \ \ln(k_m / k_a) \qquad (6.56)$$

In all such cases, it should be remembered that the concentrations referred to in the above expressions are those at the mineral surface and that these values can differ substantially from the bulk values which are reflected in the value measured as $E_H$. This is particularly true for the concentration of Fe(II) which can be significantly higher at the mineral surface where it is produced by reduction of Fe(III) and, in the case of minerals containing iron, also by anodic oxidation of the mineral. In order to obtain approximate solution potentials at the mineral surface, ring-disk electrodes of several minerals (Table 6.4) were used. In these electrodes, a platinum ring electrode surrounds (but is insulated from) the mineral disk. Thus, the potential of the ring electrode is important as an experimental indicator of the value of $E_H$ in the region close to the disk electrode. These electrodes were rotated in one of three solutions.

**Table 6.4 Summary of the potentials (versus SHE) after 1 h at 1000 rev/min.**

| Mineral | Solution | $E_H$ (V) | $E_m$ (V) | $E_H - E_m$ | $E_R$ (V) | $E_R - E_m$ (V) | |
|---|---|---|---|---|---|---|---|
| | | | | | | Obs. | Calc. |
| $FeS_2$ | (A) | 1.280 | 0.860 | 0.460 | 1.065 | 0.205 | 0.216 |
| | (B) | 0.860 | 0.829 | 0.035 | 0.852 | 0.023 | 0.020 |
| | (C) | 0.653 | 0.653 | 0 | 0.653 | 0 | 0 |
| $CuFeS_2$ | (A) | 1.325 | 0.805 | 0.520 | 1.23 | 0.425 | 0.305 |
| | (B) | 0.840 | 0.755 | 0.085 | 0.830 | 0.075 | 0.064 |
| | (C) | 0.668 | 0.660 | 0.008 | 0.668 | 0.008 | 0.001 |
| FeAsS | (A) | 1.340 | 0.73 | 0.610 | 1.10 | 0.370 | 0.331 |
| | (B) | 0.865 | 0.715 | 0.150 | 0.860 | 0.145 | 0.093 |
| | (C) | 0.668 | 0.667 | 0.001 | 0.668 | 0.001 | 0.001 |

*Data: Nicol, M.J., Lazaro, I, 2002. The role of Eh measurements in the interpretation of the kinetics and mechanisms of the oxidation and leaching of sulfide minerals. Hydrometallurgy 63, 15–22.*

**(A)** 10 g/L iron(III) as the sulphate. Residual iron(II) in this solution was 'titrated' by the addition of a solution of permanganate until the potential of a platinum electrode increased above 1.0 V versus SHE.

**(B)** 10 g/L iron(II) + 5 mg/L iron(II) as the sulphates.

**(C)** 5 g/L iron(III) + 5 g/L iron(II) as the sulphates.

Table 6.4 summarises the various potential measurements taken after one hour for each of the three mineral electrodes.

It is apparent that, as predicted, the potential of the mineral surface is almost identical to that of the bulk solution (as measured by $E_H$) and to that of the ring electrode only for solutions containing appreciable concentrations of iron(II). On the other hand, for solutions containing very low concentrations of iron(II), the difference $E_H - E_m$ can be as much as 0.5 V for all minerals. It should be noted that the ring electrode potential lies, as expected, between $E_H$ and $E_m$. The last two columns in the Table compare the experimental values of the difference $E_R - E_m$ with those that have been estimated using Eq. (6.55) above. The results are surprisingly close considering the approximations inherent in the treatment and the experimental data and lend a degree of quantitative credence to previous qualitative observations and interpretations.

### 6.4.1.2 Application of the mixed potential model to the cyanidation reaction

As discussed in a previous section, the cyanidation process involves the coupled electrochemical reactions involving the anodic oxidation of gold in the presence of the cyanide ion and the cathodic reduction of oxygen in a typical Type IV system. Both anodic and cathodic reactions are subject to mass transport limitations.

There are two limiting cases, depending on the relative concentrations of free cyanide and dissolved oxygen. These are shown in Fig. 6.14.

Case 1. $[CN^-] \gg [O_2]$

Case 2. $[O_2] \gg [CN^-]$

Thus, depending on the concentrations of cyanide and dissolved oxygen, the rate can be controlled by either cyanide or oxygen diffusion. The maximum rate of the anodic reaction (gold dissolved per unit surface area) will be given by,

$$R_A = k \ D_{CN}[CN^-]/2 \qquad (6.57)$$

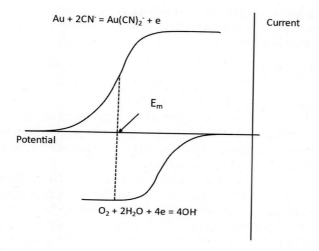

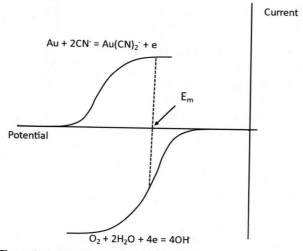

**Figure 6.14** Mixed potential model for the cyanidation of gold.

where $[CN^-]$ is the cyanide concentration, k is a mass transfer co-efficient and $D_{CN}$ the diffusion coefficient of the $CN^-$ ion.

Similarly, the maximum rate of the cathodic reaction is

$$R_C = 4 \cdot k \ D_o[O_2] \qquad (6.58)$$

Why does the expression for $R_A$ have a '2' in the denominator and the expression for $R_C$ a factor of '4'?

For a given oxygen concentration, there is an optimum concentration for cyanide which is given by the point for which $R_A = R_C$ that is when

$$[CN^-] = 8[O_2] \cdot D_O/D_{CN} \approx 8 \cdot [O_2], \quad \text{for } D_o \approx D_{CN} \qquad (6.59)$$

In air-saturated water $[O_2] = 2.5 \times 10^{-4}$ mol/L = 8 ppm.
Optimum $[CN] = 8 \times 2.5 \times 10^{-4}$ mol/L.

On the plants $[CN^-]$ is generally expressed as **ppm NaCN** which, for this case is 100 ppm. In oxygen saturated water, the dissolved oxygen concentration is approx. $5\times$ that for air, so that

Optimum $[NaCN] = 500$ ppm.

This principle can readily be demonstrated electrochemically by applying an anodic potential to oxidise a rotating gold electrode immersed in an alkaline cyanide solution in the absence of oxygen. As shown in Fig. 6.15, the rate of anodic oxidation, as measured by the anodic current, is related to the rotation speed of the electrode.

The rate of dissolution of a metal from a rotating disc can be expressed by the Levich equation:

$$J = 0.62 \ D^{2/3}\omega^{1/2}\eta^{-1/6}C \qquad (6.60)$$

in which J = flux of cyanide (mol m$^{-2}$ s$^{-1}$)
    D = diffusion coefficient of cyanide (m$^2$ s$^{-1}$)
    C = concentration of cyanide (mol m$^{-3}$)

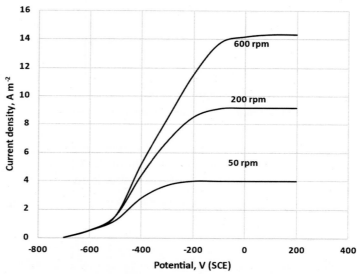

**Figure 6.15** Anodic polarisation of 2% w/w Ag/Au alloy in a solution of 140 ppm CN$^-$.

$\omega$ = rotation rate of disc $(s^{-1})$
$\eta$ = kinematic viscosity of medium $(m^2\ s^{-1})$
Note that the potential in this case is measured against the saturated calomel reference electrode.

How would you analyse the limiting current densities in Fig. 6.12 in terms of the Levich equation to calculate the diffusion coefficient of cyanide ions?

The remainder of this chapter will be devoted to a number of leaching systems that are practiced in industry. The focus will be on the chemistry of the reactions involved and, where appropriate, the mechanisms of the dissolution processes. The more practical details of leaching systems will be dealt with in Chapter 8.

## 6.5 Leaching of oxide minerals

The term 'oxide' minerals is generally used to include carbonate and silicate minerals that can be dissolved by relatively simple acid-base reactions such as (for a divalent metal M)

$$MO + H^+ = M^{2+} + H_2O \qquad (6.61)$$

$$MCO_3 + 2H^+ = M^{2+} + CO_2 + H_2O \qquad (6.62)$$

$$MSiO_3 + 2H^+ = M^{2+} + SiO_2 + H_2O \qquad (6.63)$$

In many cases, as we shall see, the minerals are more complex but the same general reactions can be written. Some oxides, such as those of uranium can only be effectively dissolved in oxidative processes and some such as those of manganese, require reductive processes. We shall deal with a number of the more important 'oxide' minerals.

### 6.5.1 Zinc oxides and silicates

The most important zinc oxide is that formed by the roasting of zinc sulphide concentrates. During the roasting process, most of the zinc is converted to ZnO but a relatively small fraction is converted to less soluble zinc ferrites such as $ZnFe_2O_4$ ($ZnO \cdot Fe_2O_3$). Thus, during the leaching of zinc calcines, the zinc oxide dissolves readily at pH values of 4–5 and temperatures of 50°C. However, the zinc ferrites require more concentrated acid

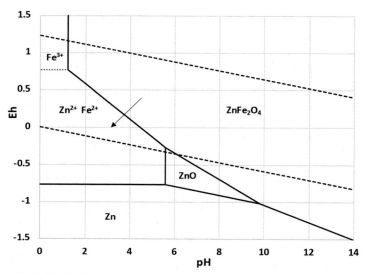

**Figure 6.16** Eh-pH diagram for the zinc-iron system at 25°C. Unit activities of soluble species.

and higher temperatures for effective leaching that is therefore carried out in several counter-current stages (see Chapter 8). The Eh-pH diagram for the zinc-iron system is shown in Fig. 6.16. Thus, dissolution of ZnO by acid simply requires moving horizontally to the left of the vertical line at pH 5.5 into the area of stability of the zinc ion. On the other hand, dissolution of the ferrite by acid requires moving horizontally to the left of the vertical line at pH 1.2 into the area of stability of the zinc ion. The iron in the ferrite would dissolve as iron(III). It is interesting to note that reductive dissolution of the ferrite is also possible as shown by the arrow in the diagram.

$$ZnFe_2O_4 + 8H^+ = Zn^{2+} + 2Fe^{2+} + 4H_2O + 6e \tag{6.64}$$

In this case, the iron dissolves as iron(II).

> On the basis of the Eh-pH diagram, suggest a possible reducing agent for zinc ferrite and an appropriate pH. Write a balanced chemical reaction for the reductive leaching reaction.

The data in Table 6.5 compares the rates of dissolution of several ferrites (relative to that for $Fe_3O_4$) in 1M HCl solutions

**Table 6.5 Comparison of the rates of dissolution of ferrites (relative to that for Fe₃O₄) in 1M HCl solution.**

| Ferrite | 1M HCl | 1M HCl + 0.01M Cu(I) |
|---------|--------|----------------------|
| $Fe_3O_4$ | 1 | 40 |
| $CuFe_2O_4$ | 0.7 | 100 |
| $ZnFe_2O_4$ | 0.08 | 50 |
| $NiFe_2O_4$ | 0.02 | 4 |

*Data: Lu, Z., Muir, D.M., 1988. Dissolution of metal ferrites and iron oxides by HCl under oxidising and reducing conditions. Hydrometallurgy 21, 9–21.*

under normal and reducing conditions (addition of 0.01M copper(I) as the reductant).

These data demonstrate that reductive leaching of ferrites is significantly faster (by at least an order of magnitude) than simple acid dissolution for all ferrites studied.

There are several important non-sulphide zinc minerals that are processed for the recovery of zinc. The most important of these are

| Willemite | $Zn_2SiO_4$ |
|-----------|-------------|
| Hemimorphite | $Zn_2SiO_3(OH)_2$ |
| Smithsonite | $ZnCO_3$ |
| Hydrozincite | $Zn_5(OH)_6(CO_3)_2$ |

The leaching reactions for these minerals are simply acid-base reactions and, because zinc is amphoteric, either acid or base can be used. Thus, the reaction for willemite with acid is

$$Zn_2SiO_4 + 2H_2SO_4 = 2ZnSO_4 + Si(OH)_4 \qquad (6.65)$$

In this and other silicate minerals, reaction with acid initially forms supersaturated solutions of silicic acid as shown in Eq. (6.65). However, this polymerises to form hydrated silica. If this polymerisation reaction is rapid, the silica can form a product layer around the dissolving mineral and inhibit the rate of dissolution. Depending on the nature of the precipitated silica, solid-liquid separation problems can be encountered when silicate minerals are leached.

The Eh-pH diagram for zinc that includes willemite is shown in Fig. 6.17.

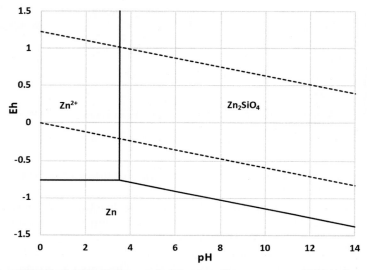

**Figure 6.17** Partial Eh-pH diagram for the zinc-silicon system at 25°C. Unit activities of soluble species.

Dissolution in acid will necessitate a reduction in the pH to values below the vertical line at pH 3.5. This can be compared to the dissolution of zinc oxide at pH 5.5 as shown in Fig. 6.16. Thus, leaching of willemite requires a lower pH than zinc oxide.

## 6.5.2 Copper oxides and silicates

Oxide, silicate and carbonate minerals of copper such as malachite, atacamite, brochantite, chrysocolla and tenorite can be leached in acid or ammoniacal solutions. We will use this group of minerals to illustrate general trends and to demonstrate how leaching data can be interpreted in terms of the rate-determining step in specific cases.

### 6.5.2.1 Acid leaching

As could be expected, the rates of dissolution of these minerals in acid solutions generally increases with increasing acidity and some data for rotating disks of tenorite (CuO) is shown in Fig. 6.18.

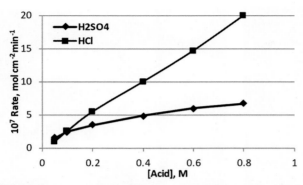

**Figure 6.18** Effect of acid concentration on the rate of dissolution of CuO in hydrochloric and sulphuric acid solutions at 30°C. Data: Majima et al., 1980.

Note that the rate is a linear function of the hydrochloric acid concentration but appears to reach a maximum at sulphuric acid concentrations approaching 1.0M. Note also that the rate is approximately the same for both lixiviants at low concentrations but is greater in hydrochloric acid at concentrations above about 0.1M. The non-linear dependence in the case of sulphuric acid has been interpreted in terms of a decreasing activity of the proton with increasing concentration. The activation energy is about 55 kJ/mol for sulphuric acid and 52 kJ/mol for hydrochloric acid dissolution. These relatively high values indicate that the rate is controlled by the rate of the chemical reaction on the surface of the mineral.

In the case of chrysocolla, a complex copper silicate mineral with the formula $(Cu,Al)_2H_2Si_2O_5(OH)_4 \cdot nH_2O$ dissolution in acid produces a silica product layer that could be expected to reduce the rate as the layer thickens. As shown by the micrographs in Fig. 6.19, the particles retain their shape after dissolution of copper (green-blue) in acid solution. In this case, it appears that rate will be determined by the diffusion of copper out of the particles.

Kinetic data for the leaching of copper is given for one particle size in Fig. 6.20. The dissolution is relatively rapid with almost complete dissolution of copper in about 1h under these conditions. A plot of Eq. (6.37) is also shown and the good linearity up to 80% dissolution confirms rate limiting diffusion of copper ions out of the silica matrix.

**Figure 6.19** Groups of chrysocolla particles before and after leaching in sulphuric acid solutions (Nicol and Akilan, 2018).

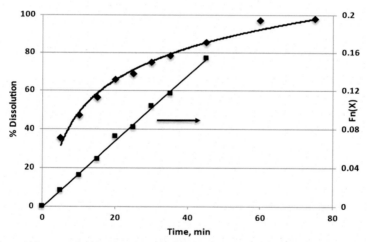

**Figure 6.20** Dissolution of chrysocolla ($+250-355\mu m$) in 2M $H_2SO_4$ at 30°C. Also shown is a plot of the function for mass transport through a product layer. Data: Nicol and Akilan, 2018.

The leaching of malachite is similar to chrysocolla in acid solutions although the absence of an insoluble product is advantageous. The overall reaction is

$$Cu_3(OH)_2(CO_3)_2 + 6H^+ = 3Cu^{2+} + 2CO_2 + 4H_2O \qquad (6.66)$$

Typical dissolution data is shown in Fig. 6.21 for two size fractions in 0.1M sulphuric acid at 30°C.

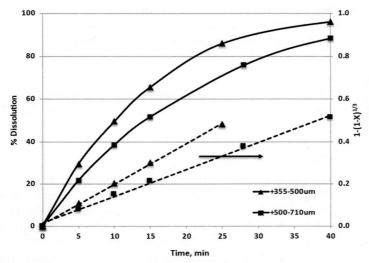

**Figure 6.21** Dissolution of malachite in 0.1M $H_2SO_4$ at 30°C. Also shown is a plot of the function for a shrinking particle. Data: Nicol, 2018.

In this case, one could expect that the rate is controlled by the rate of the chemical reaction on the surface and that the data should conform to Eq. (6.28). Thus, a plot of $1-(1-X)^{1/3}$ versus time should be linear (X is the fraction of copper dissolved) and this is the case as shown in Fig. 6.20.

Confirmation that the rate is controlled by the rate of the chemical reaction at the mineral surface is given by the following additional results.

1. The activation energy obtained from the rate constants derived from plots such as those in Fig. 6.20 at different temperatures was found to be 53.6 kJ mol$^{-1}$. This is indicative of a rate controlling chemical reaction.

2. A plot of the rate constants for different particle sizes is linear (Table 6.2).

3. The rate constants obtained from plots such as that in Fig. 6.20 are about three orders of magnitude lower that calculated for mass transport of protons to the surface of the particles using the correlation given in Chapter 5.

The rate increased with increasing acidity but the plot was only linear if the proton activity (as calculated from the measured pH) was used instead of the acid concentration showing that the rate is first order in the proton activity.

Most copper oxide minerals can also be leached in ammoniacal solutions at pH values from about 8 to 10.5 given the favourable thermodynamics due to the relatively strong complexes of ammonia with copper(II) ions (Fig. 6.4). However, this alternative has not found application in the industry.

Write balanced chemical reactions for the dissolution of malachite and chrysocolla (assume $CuSiO_3$) in solutions of ammonia and ammonium ions.

### 6.5.3 Bauxite

The Bayer process for the production of alumina from bauxite is a refining process, producing pure (>99%) alumina from impure ores (30−60% $Al_2O_3$). The process was patented in 1888, and there have since been few fundamental changes. Aluminium hydroxides in the form of the minerals gibbsite and boehmite are amphoteric and alkaline processing is preferred for several reasons, the most important of which is better control of impurities, particularly iron.

The basic reaction is the equilibrium

$$Al(OH)_3(s) + OH^- \Leftrightarrow Al(OH)_4^- \tag{6.67}$$

Dissolution takes place by reaction of the ore with concentrated caustic soda solutions at elevated temperatures in an autoclave. The solubility of aluminium ions in caustic soda is very dependent on the temperature and the NaOH concentration as shown by the data in Fig. 6.22.

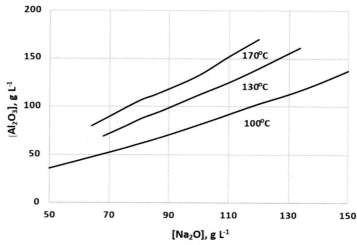

**Figure 6.22** Solubility of gibbsite in NaOH solutions. Data: Russel et al., 1955.

Thus, hot solutions from the autoclave are cooled to crystallise hydrated alumina ($Al_2O_3 \cdot 3H_2O$) and the spent solution recycled to the autoclave to dissolve more aluminium at the higher temperature. Note that at higher temperatures, there is a phase transition from gibbsite to boehmite as the stable solid phase. Ores containing gibbsite can be processed at temperatures in the range 100–150°C while those containing boehmite require temperatures above 150°C.

Bauxite dissolution is usually carried out in stirred autoclaves combined with a system of heaters and flash tanks to allow for efficient heating and cooling of the slurry. Residence times are generally 5–30 minutes. The rate limiting step is usually not the dissolution of the aluminium hydroxides but rather the dissolution and subsequent precipitation of silicate minerals. Silica, in particular in the ore, dissolves readily and can result in excessive consumption of NaOH.

## 6.5.4 Uranium minerals

Uranium mineralogy is generally quite complex with a number of minerals containing uranium in oxidation states +4 and +6. The most common U(IV) mineral is uraninite ($UO_2$) while the more refractory mineral coffinite $U(SiO_4)_{1-x}(OH)_{4x}$ is often also found. The U(VI) minerals range from the simple yet rare schoepite ($UO_3 \cdot 2H_2O$) to complex oxides such as the relatively common and refractory mineral, brannerite (U,Ce,Ca)(Fe,Ti)$_2O_6$, silicates, phosphates, aresenates and vanadates such as carnotite ($K_2(UO_2)_2(VO_4)_2 \cdot 2H_2O$). The U(VI) minerals can be leached by adding acid,

$$UO_3 \cdot 2H_2O + 2H^+ = UO_2^{2+} + 3H_2O \qquad (6.68)$$

As the Eh-pH diagram in Fig. 6.23 shows, this can be accomplished at pH values below about 4.5. As described in Chapter 3, the U(IV) minerals must be oxidatively leached, that is follow the path of the arrow in Fig. 6.23. This diagram is modified from that in Chapter 3 by inclusion of sulphate that complexes with the uranyl ion as shown. This slightly reduces the potential required to oxidise uraninite to about 0.3 V and an oxidising agent with a potential greater than this value is required.

In industry, ferric ions are used in conjunction with other oxidants such as manganese dioxide or sodium chlorate that re-oxidise the ferrous ions to ferric. In this case, the Fe(III)/Fe(II) couple acts as a **redox mediator** between the oxidant and the uraninite.

$$UO_2 + 2Fe^{3+} = UO_2^{2+} + 2Fe^{2+} \qquad (6.69)$$

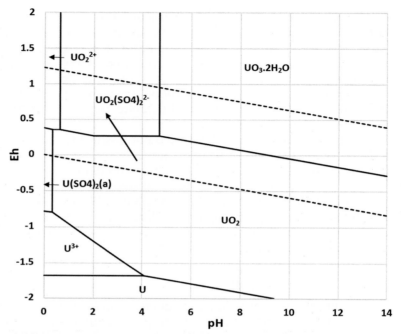

**Figure 6.23** $E_H$/pH diagram for uranium (0.002M) in the presence of sulphate(0.2M) at 25°C.

$$MnO_2 + 2Fe^{2+} + 4H^+ = Mn^{2+} + 2Fe^{3+} + 2H_2O \qquad (6.70)$$

$$ClO_3^- + 6Fe^{2+} + 6H^+ = Cl^- + 6Fe^{3+} + 3H_2O \qquad (6.71)$$

In terms of the mixed potential model for oxidative leaching, the schematic in Fig. 6.24 summarises the relevant current–potential curves.

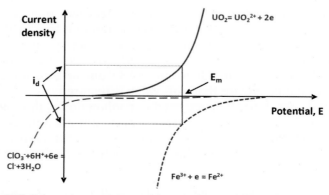

**Figure 6.24** Schematic current-potential curves for the oxidative dissolution of $UO_2$.

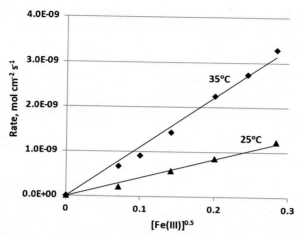

**Figure 6.25** Effect of iron(III) concentration on the rate of dissolution of $UO_2$ in 0.1M $H_2SO_4$. Data: Laxen, 1967.

Thus, reduction of ferric ions on a $UO_2$ surface is considerably more reversible that that for the reduction of chlorate (or $MnO_2$) as shown. This results in a lower mixed potential and therefore very slow oxidation of $UO_2$ by chlorate. The relatively rapid oxidation of ferrous ions by these oxidants ensures that the ratio of ferric/ferrous ions is maintained high during the leach. As shown in Chapter 5, the mixed potential model for oxidative dissolution predicts that the rate should be proportional to the square root of the oxidant concentration. The data in Fig. 6.25 for the dissolution of uraninite show a linear relationship between the measured rate and the root of the iron(III) concentration as predicted.

The uranyl ion, $UO_2^{2+}$, is unusual in that it forms relatively strong complexes with the carbonate ion as shown by the stability constant data in Table 2.4 of Chapter 2. The $E_H$/pH diagram for uranium in the presence of carbonate is shown in Fig. 3.22 of Chapter 3. Dissolution of uranium(VI) minerals can be achieved by an acid-base reaction with carbonate at pH values above about 9 by the reaction

$$UO_3 + CO_3^{2-} + 2HCO_3^- = UO_2(CO_3)_3^{4-} + H_2O \qquad (6.72)$$

In the case of uraninite, oxidation at potentials above about $-0.4$ V at pH 4 is required. In this case, the thermodynamic driving force for oxidation of $UO_2$ by dissolved oxygen is such that it can be effectively used as the oxidant. Thus, as outlined in

Chapter 8, this can be put to good use in the in situ leaching of uranium by simply pumping aerated soda water into the orebody underground.

Write a balanced chemical reaction for the dissolution of uraninite in carbonate/bicarbonate solutions using oxygen as the oxidant.

### 6.5.5 Nickel laterites

The origin of nickeliferous laterite is mostly peridotite rocks containing olivine and serpentine. Under tropical and subtropical conditions, with abundant rain fall, various acids are produced due to decay of organic materials. These acids leach out magnesium and silicon, while enriching the residue with iron and nickel. Such weathering of peridotite leads to formation of laterites. A typical nickel deposit can be divided into several distinct zones which occur at increasing depths from the surface. Table 6.6 shows typical mineralisation of nickel laterites.

**Table 6.6 Significant minerals in nickel laterites.**

|  | Formula | % Ni |
|---|---|---|
| ***Peridotite Bedrock*** | | |
| Olivine | $(Mg, Fe, Ni)_2SiO_4$ | 0.25 |
| Serpentine | $Mg_3Si_2O_5(OH)_4$ | 0.25 |
| ***Saprolite Zone*** | | |
| Nickeliferous Serpentine | $(Mg, Fe, Ni)_3Si_2O_5(OH)_4$ | 1—10 |
| 'Garnierite' | $(Ni, Mg)_3Si_4O_{10}(OH)_2$ | 10—24 |
| ***Intermediate Zone*** | | |
| Nontronite | $(Ca,Na,K)_{0.5}(Fe^{+3},Ni,Mg,Al)_4(Si,Al)_8O_{20}(OH)_4$ | 0—5 |
| ***Limonite Zone*** | | |
| Goethite | $(Fe,Al,Ni)OOH$ | 0.5—1.5 |
| 'Asbolite' | Mn, Fe, Co, Ni Oxide | 1—10 |

Metallurgical processing of lateritic nickel ore is normally carried out by one of the following.
**(i)** smelting

**(ii)** reduction roasting — ammonia leaching (Caron process)

**(iii)** sulphuric acid leaching under pressure (PAL process)

The selection of the process is usually determined by the nature of the ore. High nickel bearing limonitic ore is hydrometallurgically processed through reduction roasting — ammonia leaching, or through acid leaching. The latter is preferred when the deposit contains small quantities of acid consuming gangue.

### 6.5.5.1 Reduction roast — ammonia leaching process

In this process (Chapter 8), the ore is pre-reduced to convert the metals in the ore to an alloy containing mainly iron with nickel, cobalt and small amounts of copper that is quenched in an ammonia/ammonium carbonate solution.

The main reactions taking place during leaching of the alloy in aerated ammonia/ammonium carbonate solutions are

$$FeNi + O_2 + 8NH_3 + 2H_2O = Ni(NH_3)_6^{2+} + Fe(NH_3)_2^{2+} + 4OH^-$$

$$(6.73)$$

$$4Fe(NH_3)_2^{2+} + O_2 + 8OH^- + 2H_2O = 4Fe(OH)_3 + 8NH_3 \quad (6.74)$$

One can replace Ni in these equations with Co. The Eh/pH diagram for nickel is given in Fig. 6.26.

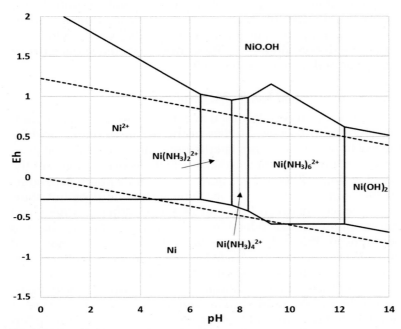

**Figure 6.26** Eh-pH diagram for the Ni—ammonia—water system. 0.1M Ni species, 2M $NH_3 + NH_4^+$.

Thus, in order to oxidise the Ni in the alloy at a pH of 10 in ammonical solution, a potential of greater than about $-0.6$ V is required. Dissolved oxygen is obviously thermodynamically capable of this oxidation with over 1 V thermodynamic driving force. However, the kinetics are slow under these conditions. Although oxygen is shown as the oxidant, research has shown that the $Co(NH_3)_6^{3+}/Co(NH_3)_6^{2+}$ couple and, to a lesser extent, the $Cu(NH_3)_4^{2+}/Cu(NH_3)_2^+$ couple is the actual oxidant. These couples play a similar role of redox mediation between dissolved oxygen and the alloy. This is possible as a result of the relatively rapid re-oxidation of the Co(II) and Cu(I) ammine complexes by dissolved oxygen.

By reference to the relevant Eh-pH diagrams in Chapter 3 demonstrate that the oxidation of Ni metal is possible using $Co(NH_3)_6^{3+}$ or $Cu(NH_3)_4^{2+}$ as the oxidant in ammoniacal solutions. Which oxidant has the more favourable thermodynamic driving force?

The Eh-pH diagram for iron in ammoniacal solutions in Chapter 3 shows a small area of stability of the $Fe(NH_3)_2^{2+}$ complex ion at pH values between 7.5 and 11. In the absence of this species, iron would simply be oxidised to an insoluble iron(III) oxide/hydroxide that would passivate the surface of the dissolving metal. The formation of an intermediate $Fe(NH_3)_2^{2+}$ allows for iron to be dissolved and to diffuse away from the surface before being oxidised by oxygen (Eq. 6.69).

### 6.5.5.2 Pressure acid leach (PAL) process

Sulphuric acid is widely used as a lixiviant in numerous hydrometallurgical processes. Studies of the use of sulphuric acid for the leaching of nickel from laterite ores under atmospheric pressure conditions has shown that for high recovery of nickel the leaching time will be long — between 3 and 24 hours depending on the temperature. Such a leaching technique does not show any selectivity between iron and nickel resulting in a high acid consumption and considerable iron in solution that will have to be removed. Primarily because of (a) high iron extraction, and (b) high acid consumption, acid leaching under atmospheric conditions has not been practiced for lateritic nickel ores.

Most of the problems associated with atmospheric acid leaching can be overcome by carrying out leaching under pressure at elevated temperatures. In order to reduce acid consumption and decrease the iron content in leach liquor, it is essential to hydrolyse the ferric ions to hematite according to the following reaction, which takes place at $\sim 200°$C.

$$2Fe^{3+} + 3H_2O = Fe_2O_3 + 6H^+ \qquad (6.75)$$

A significant fraction of the nickel in the limonitic ore (low acid-consuming MgO content) is bound in goethite-like minerals and can only be released by dissolution of these oxide minerals. The leaching process is therefore a dissolution-precipitation process which converts the iron oxides to hematite and releases the bound nickel into solution.

Manganese minerals such as asbolane, a hydrated higher oxide of manganese with substituted cobalt ions often contain most of the cobalt in laterites. In this case, addition of a small amount of a reducing agent such as elemental sulphur to the pressure autoclave assists in promoting reductive leaching of the manganese (see next section) thereby releasing cobalt and nickel. Fig. 6.27 shows typical leaching profiles for some of the more important metals during the PAL process.

The nickel recovery increases smoothly with time as the acid is consumed by the reactions. As the pH increases with time, iron will precipitate largely as goethite and hematite. The cobalt and manganese curves are similar in shape due to the presence of most of the cobalt in manganese minerals such as asbolane. The relatively high residual acid concentration is obviously not desirable as it contributes to the acid consumption and must

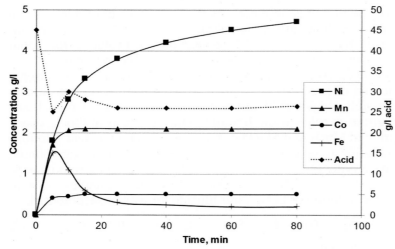

**Figure 6.27** Typical concentration profiles during a PAL process at 240°C.

be neutralised before the nickel and cobalt can be recovered. Recent developments have focused on the use of relatively reactive components in the ore such as the saprolite in a subsequent atmospheric pressure leach on the autoclave discharge. This will act to neutralise the acid while increasing nickel recovery from the saprolite component. The disadvantage is increased levels of iron in solution after the two stage leach process.

## 6.5.6 Other oxides

A number of other oxide minerals are subjected to leach processes in acidic solutions. Note that, in general compounds of the type MO. $M_2O_3$ (known as ferrites) are considerably more refractory to dissolution than the simple oxides. Thus, the dissolution in acid solutions of the minerals magnetite [$FeO \cdot Fe_2O_3$], ilmenite [$FeO \cdot TiO_2$], chromite [$FeO \cdot Cr_2O_3$], hausmannite [$MnO \cdot Mn_2O_3$], jacobsite [$MnO \cdot Fe_2O_3$], trevorite [$NiO \cdot Fe_2O_3$] and magnesioferrite [$MgO \cdot Fe_2O_3$] are all thermodynamically favourable but very slow under ambient conditions compared to the corresponding simple oxide minerals. The rate of dissolution can be considerably enhanced in the presence of a suitable reducing agent such as $SO_2$ as shown by the following reaction for oxidised manganese ores

$$MnO \cdot Mn_2O_3 + SO_2 + 4H^+ = 3Mn^{2+} + SO_4^{2-} + 2H_2O \quad (6.76)$$

Some of the more important oxide leaching systems are

### 6.5.6.1 Manganese oxides

Manganese calcines (MnO) are produced by the reductive roasting of manganese ores. As the Eh-pH diagram (Fig. 3.18) in Chapter 3 shows dissolution of MnO (or $Mn(OH)_2$) can be accomplished by the simple acid-base reaction at pH values below about 7. At this pH, iron in the calcine is not soluble. Alternative direct hydrometallurgical routes have been developed but not implemented in practice. These involve reductive dissolution of ores containing higher oxides of manganese such as manganite, hausmannite, pyrolusite and braunite. Thus, as Fig. 3.18 shows, dissolution of $MnO_2$ and MnO.OH requires reduction in acid solutions to solubilise the manganese as $Mn^{2+}$ ions. The reductant should have a potential below about 0.7 V for this to be thermodynamically possible for $MnO_2$ at a pH of 2. The most suitable reductant has been found to be sulphur dioxide for which the potential,

$$SO_4^{2-} + 4H^+ + 2e = SO_2 + 2H_2O \quad E^o = 0.158 \text{ V} \quad (6.77)$$

is adequate.

### 6.5.6.2 Titanium oxides

Ilmenite and other titanium minerals which are leached in order to produce titanyl solutions from which $TiO_2$ pigments are produced. The leaching of ilmenite in the sulphate process for the production of $TiO_2$ pigment is carried out in concentrated sulphuric acid solutions at elevated temperatures due to the slow kinetics of the reaction

$$FeTiO_3 + 4H^+ = TiO^{2+} + Fe^{2+} + 2H_2O \qquad (6.78)$$

The Eh/pH diagram for the Ti/Fe system is shown in Fig. 6.28. On the basis of this diagram, dissolution of both the titanium and the iron can be accomplished by.

**(a)** reduction in the pH to below about $-0.5$, that is concentrated acid on the basis of the above reaction.

**(b)** reductive dissolution to Ti(III) and Fe(II) at low pH values — write a balanced chemical half reaction for this reduction.

**(c)** oxidative dissolution to Ti(IV) and Fe(III)

Ilmenite is generally oxidised (weathered) with part of the iron(II) in the mineral oxidised to iron(III). This results in a more refractory (to dissolution) mineral and the following reaction is possible in which the weathered ilmenite is reduced by a suitable reducing agent — (Ti(III) ions in this case) to stoichiometric ilmenite that dissolves more rapidly

$$\{Fe^{2+} \cdot yFe^{3+} \cdot Ti^{4+} \cdot (3 + 1.5y)O^{2-}\} + 3yH^+ + yTi^{3+}$$
$$= yFe^{2+} + FeTiO_3 + 1.5yH_2O + yTi^{4+} \qquad (6.79)$$

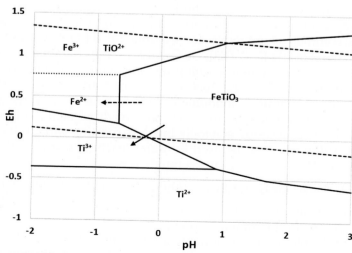

**Figure 6.28** Partial Eh/pH diagram for the titanium/iron system at 25°C (0.1M species).

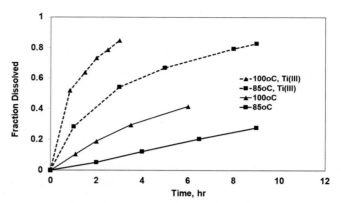

**Figure 6.29** Effect of reducing conditions (10 g L$^{-1}$ Ti(III)) on the dissolution of ilmenite in 450 g/L sulphuric acid. Data: Zhang and Nicol, 2010.

The enhanced leaching of this material in the presence of Ti(III) ions is shown in Fig. 6.29.

It is apparent that the rate can be significantly enhanced by increasing the temperature from 85 to 100°C and by conducting the leach under reducing conditions. Furthermore, the rates of dissolution of titanium and iron are similar confirming that there are no solid reaction products containing only iron or titanium.

### 6.5.6.3 Lithium silicates

Lithium is recovered from both ores (so-called hardrock) and also from brine deposits (Chapter 10). In the case of ores, the main commercial minerals are spodumene ($Li_2O \cdot Al_2O_3 \cdot 4SiO_2$) and lepidolite [(Li,K) (F,OH)]·$3SiO_2 \cdot Al_2O_3$. Both minerals are refractory to conventional leaching methods and a thermal treatment by heating to about 1000°C is required. In the case of the more commonly used spodumene, this treatment converts the mineral from the α-form into the more readily leached β-form. Subsequent processing generally involves roasting with concentrated sulphuric acid followed by water leaching of the lithium sulphate. The reaction during sulphation is

$$Li_2O \cdot Al_2O_3 \cdot 4SiO_2 + H_2SO_4 = Al_2O_3 \cdot 4SiO_2 \cdot H_2O + LiSO_4 \quad (6.80)$$

This is an ion-exchange in which protons substitute for lithium ions in the mineral structure.

The dissolution of the lithium sulphate in water is rapid. The same process can be also used for ores containing lepidolite.

## 6.6 Sulphide mattes

Matte produced by the smelting of copper/nickel sulphide concentrates can be regarded as a complex synthetic sulphide because it usually contains $Ni_3S_2$ and $Cu_{1.96}S$(djurleite), as well as minor amounts of other sulphides such as CoS, $(Ni,Fe)_9S_8$ (pentlandite), $Ni_7S_6$(godlevskite), $Cu_5FeS_4$(bornite), $Cu_5FeS_6$(i-daite) and, in some cases, PGM. There are three main processes utilised to recover Ni, Co and Cu from these sulphides.

### 6.6.1 Acid pressure leach process

This is the process used for mattes rich in PGMs and involves several stages of atmospheric and pressure leaching carried out at 80−160°C with $H_2SO_4$ to selectively leach nickel and cobalt, then copper. In the first atmospheric pressure stage, recycled $Cu^{2+}$ solution from electrowinning and the second-stage leach displaces $Ni^{2+}$ to leave an impure copper sulphide residue and NiS by the overall reaction

$$2Ni_3S_2 + 2Cu^{2+} + 0.5O_2 + 2H^+ = Cu_2S + 3NiS + 3Ni^{2+} + H_2O$$

$$(6.81)$$

This is a complex reaction that can be viewed as occurring in two parallel steps

$$Ni_3S_2 + 0.5O_2 + 2H^+ = 2NiS + Ni^{2+} + H_2O \qquad (6.82)$$

$$2Cu^{2+} + Ni_3S_2 = Cu_2S + NiS + 2Ni^{2+} \qquad (6.83)$$

While this may not be the exact mechanism, the net result is dissolution of about 50% of the nickel and effectively zero copper. The solution is used for the electrowinning of nickel after some purification.

In the second stage that is a pressure oxidation stage, the remaining Ni is dissolved together with some copper. This solution is recycled to the first stage and the residue that is copper sulphides is then oxidatively leached in a third stage under pressure at a higher temperature.

As can be appreciated, this process is one of the most complex operating plants with many recycle streams.

### 6.6.2 Ammoniacal pressure leaching

In this process, the milled matte is leached in horizontal autoclaves in the presence of ammonia and ammonium sulphate with oxygen as the ultimate oxidant.

$$Ni_3S_2 + 4.5O_2 + NH_3 + 2NH_4^+ = 3Ni(NH_3)_4^{2+} + 2SO_4^{2-} + H_2O$$

$$(6.84)$$

However, as indicated above, the Cu(II)/Cu(I) and Co(III)/Co(II) ammine couples can act as efficient redox mediators between the matte surface and the dissolved oxygen. In the absence of such species, the rate of leaching would be several orders of magnitude slower.

## 6.6.3 Chloride leaching

The Falconbridge matte leach process was the first commercially successful integrated non-ferrous metal chloride leaching process using a sulphur-rich feed.

In the original process, the matte was preferably leached with conc. HCl at 70° and was selective for $Ni_3S_2$ leaving CuS/$Cu_2S$ as a residue due to the very low solubility of the copper sulphides.

$$Ni_3S_2 + 6H^+ = 3Ni(II) + 2H_2S + H_2 \qquad (6.85)$$

Due the problems in dealing with $H_2S$, the process was modified to one in which chlorine was introduced to produce elemental sulphur. The use of chloride results in more rapid dissolution kinetics which eliminates the need for high pressure autoclaves as in Section 6.1.1. In this process, the nickel matte is leached in two stages (one under low pressure) with a solution containing $Cl_2/CuCl_2$ at 110° at a strictly controlled potential to give a solution containing 200 g $L^{-1}$ Ni, 50 g $L^{-1}$ Cu and 250 g $L^{-1}$ $Cl^-$.

$$Ni_3S_2 + 2Cu(II) = NiS + Ni(II) + 2Cu(I) \qquad (6.86)$$

$$NiS + 2Cu(II) = Ni(II) + S + 2Cu(I) \qquad (6.87)$$

$$4Cu(I) + 2Cl_2 = 4Cu(II) \qquad (6.88)$$

CuS/$Cu_2S$ reacts more slowly (see next Section) and is only partly leached. All the copper in solution is then cemented onto fresh matte and ultimately recovered as a CuS/$Cu_2S$ residue when the redox potential of the solution is decreased.

$$Ni_3S_2 + 2Cu(I) = NiS + Cu_2S + Ni(II) \qquad (6.89)$$

The CuS/$Cu_2S$ residue is roasted to CuO + $SO_2$ and leached with spent electrolyte from copper electrowinning.

This process is an excellent example of the integration of both sulphate processing (for Cu) and chloride (for Ni and Co).

## 6.7 Sulphide minerals

### 6.7.1 Copper sulphides

The thermodynamics of the dissolution of the main copper sulphide minerals chalcocite and covellite was outlined extensively in Section 6.2.3. The following are results of the dissolution of synthetic (natural samples of these minerals seldom contain single minerals) covellite, chalcocite and digenite ($Cu_{1.8}S$). In each case, the solution potential was controlled to different values by the injection of oxygen.

The results for covellite in terms of copper dissolution at the three potentials are shown in Fig. 6.30. As the thermodynamic data in Fig. 6.5 shows, dissolution is only possible at potentials above about 0.6 V in dilute chloride solutions. The results in Fig. 6.30 confirm this in that the extent of dissolution is significantly lower at a potential of 0.55 V but approximately the same at 0.60 and 0.65 V. In this case, the $E_h$ (used to control the potential) will be approximately the same as $E_m$ as can be predicted from considerations given in Section 6.4.2.1.

The results for chalcocite are summarised in Fig. 6.31.

It is apparent that the rate of copper dissolution from chalcocite is rapid compared with that of covellite with almost complete dissolution at the higher potentials after 200 h compared to 80%

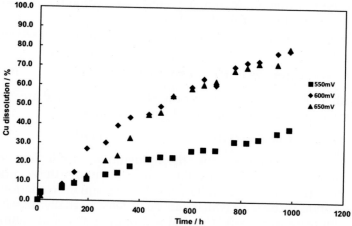

**Figure 6.30** Copper dissolution from synthetic covellite in 0.2M HCl with 0.2 g $L^{-1}$ copper(II) and 2 g $L^{-1}$ iron at 35°C at different solution potentials. Data: Miki et al., 2011.

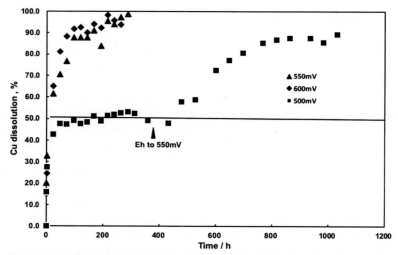

**Figure 6.31** Copper dissolution from synthetic chalcocite in 0.2M HCl with 0.2 g L$^{-1}$ copper(II) and 2 g L$^{-1}$ iron at 35 °C at different solution potentials. Data: Miki et al., 2011.

dissolution of covellite after 1000 h. At a potential of 500 mV, copper dissolution ceased after about 50 % of the copper had been dissolved but then increased at a slower rate when the potential was increased to 550 mV after about 400 h.

These results confirm that oxidation of chalcocite to covellite is rapid

$$Cu_2S + 2Fe(III) = CuS + Cu(II) + 2Fe(II) \qquad (6.90)$$

with 100% conversion after only 50 h. The intermediate 'CuS' mineral can only be dissolved at higher potentials.

$$CuS + 2Fe(III) = Cu(II) + S + 2Fe(II) \qquad (6.91)$$

This result confirms the thermodynamic prediction that oxidation of chalcocite to covellite is possible under these conditions but that 500 mV is too low for the subsequent oxidation of the secondary covellite which can be at least partially dissolved by an increase to 550 mV. Comparison of the rates of dissolution of chalcocite from 50 to 100 % dissolution in Fig. 6.31 with that for synthetic (primary) covellite in Fig. 6.30, shows that the secondary material dissolves very much more rapidly (at least an order of magnitude) than primary covellite. This effect has been ascribed to the significantly increased surface area of a porous secondary covellite formed by dissolution of approximately 50% of the copper from chalcocite.

Similar results are obtained with digenite in that the plateau for dissolution occurs at 45% copper as predicted by the stoichiometry of the reaction

$$Cu_{1.8}S + 1.6 \ Fe(III) = CuS + 0.8 \ Cu(II) + 1.6 \ Fe(II) \qquad (6.92)$$

These results show the importance of an understanding of both the thermodynamics and kinetics of a system in development of a leaching process.

## 6.7.2 Chalcopyrite

One of the major outstanding problems in hydrometallurgy is the effective leaching of chalcopyrite, particularly under ambient conditions. This sulphide mineral is one of the major sources of copper but has proved to be very difficult to leach despite a considerable amount of research over many years. The main problem appears to be the formation of a passivating layer on the surface of the mineral under oxidising conditions. This is illustrated by the current/potential curve in Fig. 6.32 which shows the anodic behaviour of the mineral in acidic solutions. The peak at about 0.6 V is associated with the formation of this layer of unknown composition.

Only uneconomically strong oxidants such as nitric acid or ozone can force the potential into the region above 0.8 V where

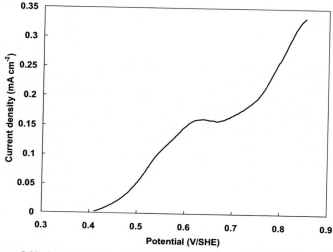

**Figure 6.32** Anodic characteristics of chalcopyrite at 25°C in 0.1M $H_2SO_4$.

relatively rapid leaching could occur while the use of the commonly available oxidants such as ferric ions or dissolved oxygen cannot produce mixed potentials on the mineral surface above about 0.7 V, that is in the passive region.

Thus, oxidation of the mineral results in an initial relatively rapid reaction which then slows considerably as shown by the potentiostatic current/time transients in Fig. 6.33 for three sulphide minerals at potentials close to the mixed potentials in acidic solutions containing iron(III) as the oxidant at ambient temperatures.

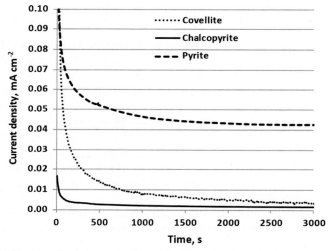

**Figure 6.33** Potentiostatic current/time transients for the oxidation of covellite, pyrite and chalcopyrite at potentials close to the mixed potentials in acidic ferric solutions at 25°C.

Thus, in the case of pyrite, the initial high current decays to a steady-state value that suggests that the rate of oxidative leaching would be significant and this is borne out in practice. However, in the case of covellite and, particularly chalcopyrite, the currents decay to very low values that, in the case of chalcopyrite, continue to decay for several days. Note that, for a four-electron process, the rate of dissolution for a current density of 1 A/m$^2$ is $2.5 \times 10^{-6}$ mol m$^{-2}$ s$^{-1}$. Thus, passivation of chalcopyrite is a slow process and can be avoided by controlling the potential of the solution during leaching to values such that the passivation potential on the surface of the mineral is not reached.

The use of high temperatures above about 75°C can partially alleviate this problem, and this has led to the development of many pressure/elevated temperature processes some of which are discussed in Chapter 8.

The mechanism of the formation of a 'passive' layer on chalcopyrite (and possibly also covellite) has been the subject of considerable research and debate. It is now widely accepted that the slow rate of dissolution of chalcopyrite is not due to coating of the mineral surface with products such as elemental sulphur or iron oxides. It has been proposed that the retardation in dissolution is the result of formation of a sulphide layer that is less reactive than chalcopyrite. According to this theory, this sulphide layer consists of ternary sulphides that are chemically and structurally different from the chalcopyrite from which they originated. These intermediate reaction products result from solid-state transformations that, in turn, are caused by the different dissolution rates of the cations in the solid leading to a change in the chemical composition of the residual sulphide phase on the surface.

The characteristic rapid decay in the anodic current for the oxidation of chalcopyrite at potentials in the region of the mixed potential under typical leaching conditions, that is below the so-called critical potential has been interpreted in terms of a model that is very similar to that found for the de-alloying process in many binary alloys. In terms of this mechanism, a relatively rapid initial but selective dissolution of the iron from chalcopyrite results in the formation of a less reactive copper-rich surface layer. Further dissolution is inhibited by relatively slow solid-state diffusion of copper and iron through this microscopically thin layer that increases in thickness with time. Both capacitance and resistance measurements are consistent with this interpretation.

## 6.7.3 Zinc sulphide

The conventional roast-leach process for zinc sulphide concentrates (Chapter 8) has been supplemented by direct pressure leaching of concentrates. In this process, several reactions take place, the most important being

$$ZnS + 2Fe(III) = Zn^{2+} + S + 2Fe(II) \qquad (6.93)$$

$$4Fe(II) + O_2 + 4H^+ = 4Fe(III) + 2H_2O \qquad (6.94)$$

that is the iron(III)/iron(II) couple acts as a redox mediator between dissolved oxygen and the mineral. The direct oxidation of sphalerite by dissolved oxygen is very slow even at 150°C. Elemental sulphur is the predominant product with small amounts of sulphate ions depending on the conditions, mainly temperature. Although the normal mixed potential model is assumed to apply to this system, it is possible that a non-oxidative mechanism could also be involved as described by the equations

$$ZnS + 2H^+ = Zn^{2+} + H_2S \qquad (6.95)$$

$$H_2S + 2Fe(III) = S + 2Fe(II) + 2H^+ \qquad (6.96)$$

The relative proportions of reaction that occur via oxidative and non-oxidative processes depends on the purity of the zinc concentrates with those having high iron contents leaching predominantly by the direct oxidative mechanism. High acid concentrations and low iron(III) concentrations favour the non-oxidative pathway.

## 6.8 Summary

In this first of two chapters on the leaching process, the reader should now have a more complete understanding of

1. The scope and variety of chemical systems that are used in the leaching of metals from various feed materials.
2. The use of Eh-pH diagrams to predict the reactions that will occur in a particular leaching system. These predictions can extend to optimisation of the conditions for leaching and the rejection of impurity dissolution.
3. The various kinetic models that can be used to describe the dissolution of particles and the problems associated with application to real ores and concentrates.
4. The mechanisms of leaching by oxidative/reductive processes in terms of the mixed potential model and its application to a number of common leaching processes.
5. The chemistry of a number of the more important leaching systems used in the industry for both oxide and sulphide ores and concentrates.

## References

Laxen, P.A., 1967. A Kinetic Study of the Dissolution of Uraninites in Sulphuric Acid. Research in Chemical and Extraction Metallurgy. Australian Institute of Mining and Metallurgy, Melbourne, pp. 181–192.

Lu, Z., Muir, D.M., 1988. Dissolution of metal ferrites and iron oxides by HCl under oxidising and reducing conditions. Hydrometallurgy 21, 9–21.

Majima, H., Awakura, Y., Yazaki, T., Chikamori, Y., 1980. Acid dissolution of cupric oxide. Metall. Trans. A 11B, 209–214.

Miki, H., Nicol, M.J., Velásquez-Yévenes, L., 2011. The kinetics of dissolution of synthetic covellite, chalcocite and digenite in dilute chloride solutions. Hydrometallurgy 106, 321–327.

Nicol, M.J., 2018. The dissolution of malachite in acid solutions. Hydrometallurgy 177, 214–217.

Nicol, M.J., Akilan, C., 2018. The dissolution of chrysocolla in acid solutions. Hydrometallurgy 178, 7–11.

Nicol, M.J., Lazaro, I., 2002. The role of Eh measurements in the interpretation of the kinetics and mechanisms of the oxidation and leaching of sulfide minerals. Hydrometallurgy 63, 15–22.

Russel, A.S., Edwards, J.D., Taylor, C.S., 1955. Solubility and density of hydrated alumina in NaOH solutions. J. Metals. 7, 1123–1128.

Zhang, S., Nicol, M.J., 2010. Kinetics of the dissolution of ilmenite under reducing conditions. Hydrometallurgy 106, 196–204.

# Problems—leaching

1.  Predict and briefly write in a tabular form (see below) the effect of the following changes of conditions on the rate of leaching of a slurry of mineral M when stirred with a reagent R in solution, assuming the rate determining step is transport controlled (A), and chemically controlled (B). Justify your answers.

| Change of conditions (all others remain constant) | A (transport Control) | B (chemical Control) |
|---|---|---|
| **i.** regrinding the mineral | | |
| **ii.** placing baffles into the leach vessel | | |
| **iii.** adding a reagent which complexes with R (e.g. adding chloride if R=Fe(III)) | | |
| **iv.** increasing the temperature from 25 to 60°C | | |
| **v.** saturation of the solution with the product of reaction (e.g. adding $CuSO_4$ to a slurry of CuO in $H_2SO_4$) | | |

2. (a)  Derive an expression which relates the extent of reaction to the time for the leaching of a mineral which consists of round, flat platelets of radius R and thickness R/100. Assume that the rate is controlled by the rate of the chemical reaction and that reaction only occurs on the planes of the platelets and not on the edges. Compare the time for

complete dissolution with that for complete dissolution of spherical particles of the same volume under the same conditions.

**(b)** Calculate the time required to completely dissolve a flat gold platelet of thickness (h) of 1μm in a cyanide solution of concentration $10^{-4}$ mol $L^{-1}$ which is saturated with oxygen ($10^{-3}$ mol $L^{-1}$) if the mass transfer coefficient (k) at the particle surface is $10^{-3}$ cm $s^{-1}$ and the density of gold is 20 g $cm^{-3}$. The rate can be expressed by $-dh/dt = k[CN^-]/\rho$ where $\rho$ is the molar density of gold.

3.  Calculate the time required to completely dissolve a 50 μm (diameter) spherical gold particle suspended in an aerated solution containing an excess of cyanide. The solubility of oxygen is $2.5.10^{-4}$ mol $L^{-1}$ and the mass transfer coefficient for a 50 μm gold particle is $1.10^{-3}$ cm $s^{-1}$. The density of gold is 19.3 g $cm^{-3}$.

4.  The copper mineral chrysocolla occurs in many heap leach operations. It dissolves in the acidic sulphate leach solutions according to the equation

$$CuSiO_3 \cdot 2H_2O(s) + H^+(aq) = SiO_2 \cdot 2H_2O(s) + Cu^{2+}(aq) + H_2O$$

The following data were obtained in a batch leach of 100 μm pure chrysocolla particles at 25°C at a controlled pH of 1.0.

| Time (s) | Fraction copper dissolved |
|---|---|
| 90 | 0.20 |
| 140 | 0.30 |
| 190 | 0.40 |
| 270 | 0.50 |
| 380 | 0.60 |
| 510 | 0.70 |
| 720 | 0.80 |
| 1020 | 0.90 |

**(a)** Demonstrate graphically that these data are consistent with a process whose rate is controlled by diffusion within the silica product layer formed around the particles which can be assumed to be spherical.

**(b)** What experiments would enable one to distinguish between rate-controlling diffusion of protons into the product layer or copper ions out of the product layer.

5. The following data were obtained during small-scale agitated batch tests for the reductive dissolution of pure manganese dioxide (230 μm approx. spherical particles) by a large excess of ferrous ions at a concentration of 0.1 mol dm$^{-3}$ in excess sulphuric acid at a concentration of 0.05 mol L$^{-3}$ at 50°C.

| Time (h) | % Dissolution |
|---|---|
| 0.6 | 24 |
| 1.2 | 42 |
| 1.8 | 56 |
| 2.4 | 67 |
| 3.0 | 76 |
| 3.6 | 83 |
| 4.2 | 87 |
| 4.8 | 92 |

The concentrations of $Fe^{2+}$ and $H^+$ and the average particle size were each varied while keeping other parameters constant. The following data were obtained for the half-life of the particles (time for 50% dissolution) at 50°C.

| Particle size ($<$μm) | $[Fe^{2+}]$ (mol/dm$^3$) | $[H^+]$ (mol/dm$^3$) | $t_{0.5}$ (h) |
|---|---|---|---|
| 42 | 0.1 | 0.05 | 0.37 |
| 57 | 0.1 | 0.05 | 0.52 |
| 116 | 0.1 | 0.05 | 1.01 |
| 230 | 0.1 | 0.05 | 2.12 |
| 48 | 0.05 | 0.07 | 0.43 |
| 48 | 0.10 | 0.07 | 0.30 |
| 48 | 0.20 | 0.07 | 0.23 |
| 48 | 0.25 | 0.07 | 0.21 |
| 48 | 0.05 | 0.07 | 0.40 |
| 48 | 0.05 | 0.05 | 0.49 |
| 48 | 0.05 | 0.03 | 0.62 |
| 48 | 0.05 | 0.01 | 1.13 |

(a) Write a balanced equation for the reaction.
(b) Show by suitable graphical or other means that the above data are consistent with a dissolution process whose rate is controlled by chemical reaction and not mass transport.

    **(c)** From the dependence of the rate on the concentrations of $Fe^{2+}$ and $H^+$, derive a rate equation which takes into account the particle size and the reactant concentrations.

    **(d)** Propose by way of appropriate current–potential curves a mixed-potential electrochemical model for the dissolution reaction. What can you infer, with reasons, about the rate-determining step in the cathodic half-reaction from the above data?

**6.** The rate of the oxidative leaching of covellite(CuS) by oxygen in acidic solutions is slow and therefore has to be carried out at elevated temperatures($>150°C$) and pressures. It has been found that the presence of dissolved iron catalyses this reaction. By means of schematic current–potential curves, suggest a mixed-potential model for the catalysed and uncatalysed reactions. What conditions in terms of relative rates are necessary for catalysis to be observed?

**7.** When finely divided gold is leached in HCl and chlorine at $25°C$, the rate is controlled by the diffusion of chlorine. When 50 g of gold consisting of 10 micron spherical particles were leached in 5 L of 1M HCl, 50 ppm $Cl_2$ (1 atmosphere) and $10\ g\ L^{-1}$ of gold, the **initial** leaching rate was 55 g gold $h^{-1}$.

    **(a)** Write a balanced equation for the reaction and schematically illustrate the mixed-potential model appropriate to this system.

    **(b)** Calculate the equilibrium potentials for the anodic and cathodic reactions and the mixed potential for the leaching system.

    **(c)** Calculate the time required for the complete leaching of 50 micron particles. State your assumptions.

$$\text{Data}: Au^{3+} + 3e = Au \quad E^o = 1.45V$$

$$Cl_2(g) + 2e = 2Cl^- \quad E^o = 1.36V$$

$$Au^{3+} + 4Cl^- = AuCl_4^- \quad \beta 4 = 2.10^{23}$$

    The Tafel slope for the anodic dissolution of gold in 1M HCl is 30 mV/decade and the exchange current density is 1 mA cm$^{-2}$.

**8.** Pyrrhotite is leached in ferric sulphate solutions according to the reaction

$$FeS + 2Fe^{3+} = 3Fe^{2+} + S$$

It has been found that the leaching rate is proportional to the square root of the iron(III) concentration.

(a) Derive a model based on the electrochemical mechanism of leaching to explain these kinetics.

(b) The leaching can be approximated to a shrinking particle model. Show that this can be expressed in the form

$$dX/dt = k[Fe^{3+}]^{0.5}(1-X)^{2/3}$$

where X is the fraction of the particle leached. Derive an expression for k.

9. The leaching of sphalerite (zinc sulphide) by ferric ions has been found to yield elemental sulphur as the product of oxidation of the sulphide ion.

(a) Write a balanced chemical reaction for the leaching process.

(b) A plot of the function $1-3(1-X)^{2/3} + 2(1-X)$ (where X is the fraction reacted) versus time was found to be linear for 50 μm diameter particles. On the basis of this, suggest the rate-limiting step for the leaching reaction.

(c) If the time for complete dissolution of the 50 μm particles is 2 h, what will be the time for complete dissolution of 100 μm particles?

(d) What would be the expected dependence of the rate on the concentration of ferric ions?

(e) Sketch the current potential curves which will be appropriate for this process indicating the position of the mixed potential.

10. The dissolution of spherical copper particles in aerated ammoniacal solutions produces the soluble $Cu(NH_3)_4^+$ ion and the rate was found to be proportional to the partial pressure of oxygen and to have an activation energy of 20 kJ/mol.

(a) Write a balanced equation for the overall reaction and schematically illustrate the mixed-potential model appropriate to this system.

(b) Calculate the **initial** leaching rate (in g copper/min) when 50 g of 100 micron diameter particles are leached in 1 L of a solution saturated with air and containing excess ammonia. Estimate the initial mass transfer coefficient from Stokes' Law given that the diffusion coefficient of oxygen is $1 \times 10^{-5}$ cm$^2$ s$^{-1}$

(c) Calculate the time required for **50%** dissolution of the copper particles.

(d) Show by way of schematic current/potential curves how the rate of dissolution and the mixed potential will change if (i) the solution is saturated with oxygen instead of air,

(ii) the agitation is increased (iii) the ammonia concentration is decreased, (iv) the particle size is decreased

**Data**: $Cu^{2+} + 2e = Cu \quad E^\circ = 0.35V$

$Cu^{2+} + 4NH_3 = Cu(NH3)_4^{2+} \quad \beta_4 = 10^{12}$

Density of copper $= 8.9$ g cm$^{-3}$

The solubility of oxygen in an ammoniacal solution saturated with air is $2.5 \times 10^{-4}$ mol L$^{-1}$

# Case study 1

Zinc ferrite of the general formula $(Zn_{1-x}, Fe_x^{2+})Fe_2^{3+}O_4$ (x<0.4) is the principal constituent of the 'neutral leach residues' which are generally obtained after the leaching of zinc calcines in spent electrolyte at pH values in the range of 4–6. In order to recover additional zinc from these residues, a so-called hot acid leach is often conducted on these materials. The data shown below were obtained during an extended investigation of the leaching of actual neutral leach residues in a number of batch leaching tests.

The dissolution reaction is

$$(Zn_{1-x}, Fe_x^{2+})Fe_2^{3+}O_4 + 8H^+ = (1-x)Zn^{2+} + xFe^+ + 2Fe^{3+}$$
$$+ 4H_2O$$

You are required to.

1. Analyse the data as thoroughly as possible in order to derive an overall rate equation which will enable you to predict the recovery of zinc as a function of time under any conditions of particle size, acid concentration, temperature and concentrations of zinc in solution.

2. Suggest a possible reason for the unusual independence of the rate on particle size in the range tested. Remember that this material has already been through a leaching process. This observation obviously makes modelling of the kinetics easier.

3. Suggest reasons for the inhibiting effect of added zinc ions on the rate.

   Unless stated otherwise, leaching was carried out on the −22+16um size fraction at 85°C in a solution of 1M $H_2SO_4$ in a mechanically agitated laboratory reactor under conditions of a large excess of lixiviant solution. The extent of leaching (X) of zinc is shown as 'Zn conv'.

**1.** Effect of agitation

| Time | Zn conv. | Zn conv. |
|---|---|---|
| (min) | 650 rpm | 1150 rpm |
| 0 | 0 | 0 |
| 30 | 0.33 | 0.35 |
| 60 | 0.6 | 0.6 |
| 90 | 0.74 | 0.75 |
| 120 | 0.83 | 0.84 |
| 150 | 0.92 | 0.91 |
| 180 | 0.98 | 0.97 |

**2.** Effect of particle size

| Time | Zn conv. | Zn conv. | Zn conv. |
|---|---|---|---|
| (min) | 9 $\mu$m | 20 $\mu$m | 28 $\mu$m |
| 0 | 0 | 0 | 0 |
| 30 | 0.37 | 0.35 | 0.32 |
| 60 | 0.62 | 0.6 | 0.55 |
| 90 | 0.73 | 0.75 | 0.72 |
| 120 | 0.81 | 0.84 | 0.83 |
| 150 | 0.9 | 0.91 | 0.92 |
| 180 | 0.96 | 0.97 | 0.96 |

**3.** Effect of acid concentration

| Time | Zn conv. | Zn conv. | Zn conv. | Zn conv. |
|---|---|---|---|---|
| (min) | 0.25M | 0.5M | 1.0M | 1.5M |
| 0 | 0 | 0 | 0 | 0 |
| 30 | 0.19 | 0.27 | 0.35 | 0.44 |
| 60 | 0.32 | 0.46 | 0.6 | 0.69 |
| 90 | 0.46 | 0.62 | 0.75 | 0.86 |
| 120 | 0.55 | 0.72 | 0.84 | 0.93 |
| 150 | 0.62 | 0.79 | 0.91 | |
| 180 | 0.69 | 0.86 | 0.97 | |
| 240 | 0.8 | | | |

**4.** Effect of temperature

| Time | Zn conv. | Zn conv. | Zn conv. | Zn conv. |
|------|----------|----------|----------|----------|
| (min) | 339K | 348K | 358K | 368K |
| 0 | 0 | 0 | 0 | 0 |
| 30 | 0.14 | 0.23 | 0.35 | 0.52 |
| 60 | 0.26 | 0.39 | 0.6 | 0.81 |
| 90 | 0.34 | 0.53 | 0.75 | 0.96 |
| 120 | 0.41 | 0.66 | 0.84 | |
| 150 | 0.47 | 0.75 | 0.91 | |
| 180 | 0.52 | 0.8 | 0.97 | |
| 240 | 0.62 | 0.9 | | |

**5.** Effect of zinc sulphate concentration

| Time | Fe conv. | Fe conv. | Fe conv. |
|------|----------|----------|----------|
| (min) | $[Zn]_o = 0$ | $[Zn]_o = 0.7M$ | $[Zn]_o = 1.5M$ |
| 0 | 0 | 0 | 0 |
| 30 | 0.35 | 0.16 | 0.13 |
| 60 | 0.6 | 0.34 | 0.29 |
| 90 | 0.75 | 0.48 | 0.41 |
| 120 | 0.84 | 0.61 | 0.52 |
| 150 | 0.91 | 0.72 | 0.61 |
| 180 | 0.97 | 0.81 | 0.68 |
| 240 | | 0.92 | 0.8 |

Note that $[Zn]_o$ is the initial added zinc sulphate. In this case, it was not possible to measure the conversion of zinc due to the high background of zinc and the iron conversion was measured.

# 7

# Theory and applications of electrochemistry in hydrometallurgy

Electrochemistry is involved, either directly or indirectly in many of the unit operations used in hydrometallurgical processes. Besides the obvious applications in the electrowinning and electrorefining of many metals and metallic compounds, there are other less obvious areas in which electrochemistry plays a fundamental role in terms of the thermodynamics and kinetics of the reactions taking place. The most important of these involve oxidative (or reductive) dissolution, cementation and chemical reduction. Examples of these applications will be dealt with in detail in the second half of the course.

In Chapter 3, the thermodynamic aspects of electrochemistry were dealt with in terms of electrode potentials. The kinetics of reactions at electrodes were summarized in Chapter 5 and included the important topic of mass transport in electrochemical reactions. In this Chapter, we will extend the theoretical aspects to include more complex reactions. The various techniques that can be applied to the study of the kinetics of electrode reactions will be reviewed and, finally, we will deal with some important practical aspects of the conduct of electrochemical research.

## 7.1 Current—potential relationships

As shown in Chapter 5, the equation for a current—potential relationship is given by the Butler—Volmer equation. This equation does not take into account mass transport of electroactive species to or from the electrode. The equation for the current-potential curve including mass transport in terms of the limiting current density $i_L$, can be shown to be the following form for a cathodic reaction under high-field conditions,

$$\eta = b \cdot \{\log(i_o / i_c) - \log[(i_L - i_c) / iL)]\} \tag{7.1}$$

Hydrometallurgy. https://doi.org/10.1016/B978-0-323-99322-7.00003-X

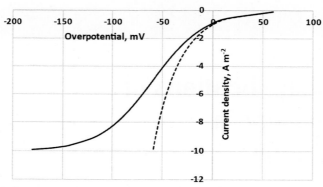

**Figure 7.1** Current-potential curve for a cathodic reaction involving mass transport.

The curve will take on the shape as shown as the black line in Fig. 7.1 for $i_L = 10\ Am^{-2}$

Note that the dotted line is the cathodic branch of the Butler–Volmer equation shown in Fig. 7.1 which does not take account of mass transport. We have ignored the reverse half-reaction by using the simplified high-field approximation.

For reactions that involve more than one electroactive species such as

$$Fe(III) + e = Fe(II)\ \text{or}$$

$$Cu(II) + e = Cu(I)$$

the situation is more complex. We can write the above in general terms as

$$Ox + ne = Red$$

Note that the actual ionic species for the metal ions has been avoided as it depends on the lixiviant system in use such as chloride or ammonia.

The kinetics of the cathodic reactions are best described by the Butler–Volmer equation (Chapter 5) which is conveniently written in the form

$$i = -nFk_f\big[[Ox]_s \exp\{-\alpha F(E - E_f)/RT\} - [Red]_s \exp\{(1-\alpha)F(E - E_f)/RT\}\big] \tag{7.2}$$

in which, using convenient units,
    i is the current density (A cm$^{-2}$).
    F is the Faraday (96480 A s mol$^{-1}$).

$k_f$ is a potential independent heterogeneous electrochemical rate constant (cm s$^{-1}$).

$[Ox]_s$ and $[Red]_s$ are the concentrations **at the electrode surface** (mol cm$^{-3}$).

$\alpha$ is the so-called transfer coefficient (normally assumed to be 0.5).

E is potential with respect to any reference electrode, V.

$E_f$ is the formal (or conditional) potential for the particular solution under study defined as the equilibrium potential at [Ox] = [Red] using the same reference electrode.

Note (1) The net negative current by convention for cathodic processes.

(2) At $E = E_f$,

$$i = -nFk_f([Ox]_s - [Red]_s) = 0 \text{ for } [Ox]_s = [Red]_s \qquad (7.3)$$

$$-nFk_f[Ox]_s = nFk_f[Red]_s = i_{o,f} \qquad (7.4)$$

in which $i_{o,f}$ is the **exchange current density** (A cm$^{-2}$) at the formal potential $E_f$.

The surface concentrations of the reacting species will not be equal to the bulk concentrations because of generation or consumption by the electrochemical reactions. Thus, using Fick's First Law (Chapter 5), one can write

$$[Ox]_s = [Ox]_o + i/nFk_L^O \qquad (7.5)$$

$$[Red]_s = [Red]_o - i/nFk_L^R \qquad (7.6)$$

in which,

$[Ox]_o$ is the bulk concentration of Ox (mol cm$^{-3}$) and

$k_L^O$ is the mass transfer coefficient (cm s$^{-1}$) for transport of Ox to the surface of the electrode.

For a rotating disk electrode, the mass transfer coefficient for Ox is given by the Levich equation

$$k_L^O = 0.620 \, v^{1/6} \, D_{Ox}^{2/3} \, w^{1/2} \qquad (7.7)$$

in which,

v is the kinematic viscosity of the solution (cm$^2$ s$^{-1}$).

$D_{Ox}$ is the diffusion coefficient of Ox (cm$^2$ s$^{-1}$).

w is the rotation speed of the electrode (radian s$^{-1}$).

Eqs. (7.5) and (7.6) can be substituted into (7.2) for the common case of n=1, to give (after some mathematical manipulation),

$$i = - \frac{(F\exp(-\alpha f(E))\{[Ox]_o - [Red]_o \exp\{f(E)\}}{\dfrac{1}{k_f} + \dfrac{\exp\{(1-\alpha)F(E)\}}{k_L^R} + \dfrac{\exp\{-\alpha f(E)\}}{k_L^O}} \qquad (7.8)$$

in which $f(E) = F(E-E_f)/RT$.

A very common case occurs when the solution only contains the Ox species, in which case, the above can be written in the following form by setting $[Red]_o = 0$,

$$1/i = 1/i_c + 1/i_{ap} + 1/i_L \qquad (7.9)$$

in which,

$$i_c = - Fk_f[Ox]_o/\exp\{\alpha f(E)\} \qquad (7.10)$$

$$i_{ap} = - F k_L^R[Ox]_o/\exp\{f(E)\} \qquad (7.11)$$

$$i_L = - F k_L^O[Ox]_o \qquad (7.12)$$

$i_C$ is the equation for the reduction of Ox in the absence of mass transport restrictions.

$i_{ap}$ is the contribution of the back reaction (oxidation of Red) to the overall current density

$i_L$ is the limiting current density for the reduction of Ox.

For the case in which $[Red]_o$ is not zero, simply replace $[Ox]_o$ in Eqs. (7.10–7.12) with

$$[Ox]_o - [Red]_o \exp\{f(E)\}$$

The relative contributions of each term to the overall current density will depend on the magnitude of each term with the smallest being the most important. The relative importance of each term becomes apparent when one considers the following two figures which summarize the results of a simulation of the Cu(II) + e = Cu(I) system on a rotating disk electrode in concentrated chloride solutions containing only $3 \text{ g L}^{-1}$ Cu(II).

Fig. 7.2 shows the calculated i/E curves for the reduction at rotation speeds of 100 and 1000rpm. In this Figure, the back reaction of oxidation of Cu(I) has been ignored, that is only $i_c$ (Eq. 7.10) and $i_L$ (Eq. 7.12) have been combined to give the overall current density at various potentials. The typical S-shaped curve is obtained with limiting current densities in each case. Note that

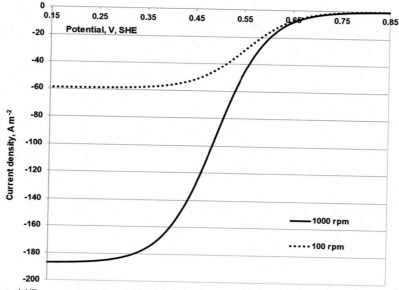

**Figure 7.2** Calculated i/E curves for the reduction of Cu(II) on a rotating Pt disk electrode from a chloride solution containing 3 g L$^{-1}$ Cu(II). The reverse reaction has been ignored.

the two curves are very similar at low current densities where the rate of mass transport ($i_L$) is large compared to the electrochemical reduction rate($i_C$). Note that for a rotating disk, the limiting current density is proportional to the square root of the rotation speed.

Of course, the reverse reaction can only be ignored in practice under certain conditions (see below) and it generally should also be included. Inclusion of $i_{ap}$ in this simulation gives the curves shown as the full and dashed lines in Fig. 7.3 in which the y-axis has been expanded. The dotted curve is the same as that for 100 rpm shown in Fig. 7.2.

It is apparent that inclusion of the reverse reaction results in a shift of the curves to more negative potentials accompanied by a greater difference between the curves at the two rotation speeds even at low overpotentials. This is a result of the simultaneous oxidation of Cu(I) produced by reduction of Cu(II) at the electrode surface. The calculated anodic current density for the oxidation of Cu(I) is also plotted in Fig. 7.3.

There are some interesting features.

1. The contribution of the reverse reaction shows a maximum anodic current density at about 0.6 V. At lower potentials, the rate of this reaction decreases as expected because it is an anodic reaction for which the rate increases as the potential

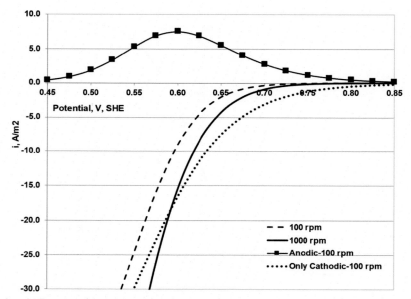

**Figure 7.3** Calculated i/E curves (with the reverse reaction) for the reduction of Cu(II) on a rotating disk electrode from a chloride solution containing 3 g $L^{-1}$ Cu(II).

increases. At higher potentials, the rate of reduction of Cu(II) decreases and therefore the concentration of Cu(I) at the electrode surface will decrease.

2. The rate of the reverse reaction is greater at the **lower** rotation speed (not shown). This is understandable given that Cu(I) formed by reduction of Cu(II) at the surface will be transported more rapidly from the electrode surface at the higher rotation speed and therefore not available for oxidation.

3. As a consequence of the magnitude of the reverse reaction being dependent on mass transport of Cu(I) from the surface, potentiostatic measurements at potentials in the region of the anodic peak in Fig. 7.3 will show a Levich-like dependence on rotation speed at current densities well below the limiting current density for reduction of Cu(II). In this case, the 'limiting' currents are due to transport of Cu(I) from the disk and not Cu(II) to the disk.

4. The shapes and relative magnitudes of the curves will depend on the magnitude of the electrochemical rate constant $k_f$ and mass transport coefficients $k_L$ (note that both

have units cm s$^{-1}$). The curves in Fig. 7.3 were drawn for $k_f = 1 \times 10^{-3}$ cm s$^{-1}$ and $k_L^{Ox} = 4 \times 10^{-3}$ cm s$^{-1}$ at 1000 rpm.

As $k_f$ increases relative to $k_L$, $i_C$ becomes increasingly less important in determining the overall current density and for $k_f \gg k_L$, one can show that

$$i = i_L / \{1 + D_R / D_O \cdot \exp(-f(E))\} \qquad (7.13)$$

This is the equation of an S-shaped curve with $i = i_L/2$ at a potential (known as the half-wave potential)

$$E_{1/2} = E_f - RT/F \ln(D_O / D_R) \sim E_f \qquad (7.14)$$

since, for most couples, $D_O \sim D_R$.

The reduction process is then said to be reversible (in a kinetic rather than thermodynamic sense). The degree of reversibility is therefore measured in terms of the relative magnitude of the rate constant $k_f$. As a general rule, the process can be classed as reversible for $k_f/k_L > 30$ and irreversible for $k_f/k_L < 0.1$.

Note that for most mineral electrodes, the reverse of the anodic dissolution reaction can be ignored, that is the anodic reaction is highly irreversible. This is not true, however, for the reactions involving the oxidants especially those involving single electron transfers such as the iron(III)/iron(II) and copper(II)/copper(I) couples.

There are other more complex reaction schemes but we will not deal with them. The above is simply an outline of how one would go about deriving the current-potential relationships for electrochemical reactions.

## 7.2 Experimental techniques

### 7.2.1 Electrodes

While simple metal electrodes such as platinum, gold and silver can be obtained commercially, most electrodes have to be fabricated from metal or mineral samples. Fig. 7.4 shows a typical disk electrode that can be simply fabricated for laboratory studies. Thus, a roughly cylindrical or cubic sample is attached to a stainless steel or brass stub using silver epoxy cement. The resulting unit is encased in epoxy resin and machined into a cylindrical electrode. It can be screwed onto the end of a shaft as part of a rotating electrode or simply used as a stationary electrode.

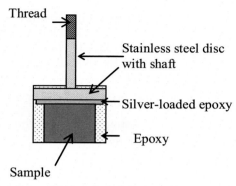

**Figure 7.4** A typical disk electrode system.

If solid mineral samples are not available, then so-called carbon paste electrodes can be used. These are prepared by mixing and grinding paraffin oil with graphite powder and the mineral powder in an agate mortar until a homogeneous paste is obtained. This paste is then packed into an insulated cavity with a metallic contact at the back. The surface can then simply be smoothed with a paper sheet. While suitable for the study of anodic processes, one has to be careful that cathodic reactions can occur on both the exposed mineral surface as well as the exposed graphite surfaces and it is not a simple process to separate the two contributions.

In most applications, a three-electrode system is used in which the material of interest is the working electrode. A counter electrode is used to complete the circuit in the cell and is composed of an inert material such a platinum, gold or graphite. Its construction is not important as long as it can pass the required currents without high voltage drops. In many cases in which one requires the electrolyte composition to be unaffected by the products of the reaction at the counter electrode, it can be housed in a glass tube with a fritted glass end piece. Finally, a suitable reference electrode is required, the position of which is important to minimize potential drops in the solution. A suitable system makes use of a Luggin capillary shown in Fig. 7.5. This enables the potential to be measured as close as possible to the working electrode surface without disturbing the current distribution to the surface. A typical setup for a rotating disk electrode is shown in Fig. 7.5.

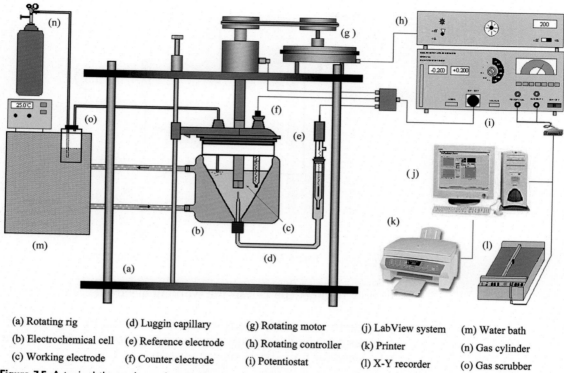

| (a) Rotating rig | (d) Luggin capillary | (g) Rotating motor | (j) LabView system | (m) Water bath |
| (b) Electrochemical cell | (e) Reference electrode | (h) Rotating controller | (k) Printer | (n) Gas cylinder |
| (c) Working electrode | (f) Counter electrode | (i) Potentiostat | (l) X-Y recorder | (o) Gas scrubber |

**Figure 7.5** A typical three-electrode rotating disk electrode setup. (a: Assembly, b: Cell, c: Working electrode (disk), d: Luggin capillary, e: Reference electrode, f: Counter electrode, g: Motor, m: Water bath, n: Gas cylinder, o: Gas bubbler).

## 7.2.2 Instruments

The workhorse of electrochemistry is the potentiostat that, as the name implies, maintains the potential of the working electrode with respect to the reference electrode at a potential that is set by the user. A simplified schematic is given in Fig. 7.6. There are several other more complex designs.

In operation, the high gain difference amplifier shown as the triangle passes current through the counter electrode/working electrode combination to maintain the potential of the reference electrode ($E_R$) with respect to the working electrode (grounded in this design) equal to the set potential ($E_i$), that is the difference $E_R - E_I = 0$. In the configuration shown, the current that flows is measured by the voltage drop across the series resistor R. Note that the role of the counter electrode is simply to allow current

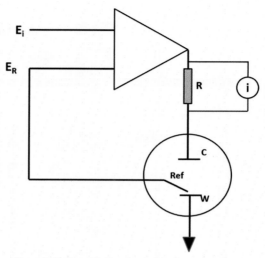

**Figure 7.6** Simplified schematic of a potentiostat.

to flow through the cell to the working electrode. The potential applied by the user to $E_i$ can, in its simplest form, be a fixed voltage. In addition, various waveforms can also be applied as $E_i$ such as a triangular voltage-time waveform for voltammetry or a sine wave for impedance measurements.

## 7.3 Common measurements

There are many available electrochemical techniques and methods and it is not possible to deal with all. Essentially, all measurements involve only the effect of potential on the current flowing at an electrode. The only difference is the nature of the potential (or current) perturbation. In the case of impedance or AC methods, in addition to measurement of the magnitude of the response to a sine wave potential perturbation, the phase angle of the resulting current response is also measured. The following are the main methods that are useful in the study of hydrometallurgical reactions.

### 7.3.1 Open circuit potential measurements

All potentiostats provide the facility of measuring the potential of the working electrode relative to any suitable reference electrode under open-circuit conditions. Of course, any high impedance digital voltmeter such as a pH meter could measure this potential. This potential is, depending on the particular application, variously called the open-circuit potential (OCP), equilibrium potential ($E_e$ or $E_H$) or the mixed potential ($E_m$). The use of OCP to describe this potential encompasses both the equilibrium and mixed potentials. In many cases, the term oxidation/reduction potential (ORP) is used in the context of the equilibrium potential. The difference is simply that the ORP is quoted as the potential relative to the reference electrode while the $E_H$ is generally defined in terms of the Standard Hydrogen Electrode (SHE). The difference between $E_H$ and $E_m$ is often not well understood, and we shall return to the distinction later in the course. An example is shown in Fig. 7.7.

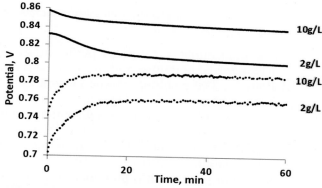

**Figure 7.7** Equilibrium (solid lines, platinum electrode) and mixed potentials (dotted lines) of a chalcopyrite electrode in chloride solutions containing 2 and 10 g $L^{-1}$ copper(II) ions.

Note that the values of $E_H$ are greater than those of $E_m$ and this is generally the case for a reacting system-in this case the oxidation of chalcopyrite by copper(II) ions. Normally, $E_H$ should be independent of time if the system is at equilibrium. In this case, the reaction produces copper(I) as a product and the copper(II)/copper(I) ratio decreases with time resulting in a decrease in the $E_H$.

Although simple in terms of measurement, these OCP's can provide useful information about the system as we shall see later.

## 7.3.2 Potentiostatic measurements

In potentiostatic or potential step measurements, the potential is instantaneously changed from one value $E_1$ (normally the OCP) to another $E_2$ as shown in Fig. 7.8.

The resulting current is then measured as a function time. For example, for the simple reaction

$$Fe(II) = Fe(III) + e$$

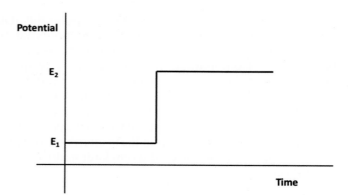

**Figure 7.8** Potential step experiment.

on, for example, a **stationary** pyrite surface, $E_1$ is the OCP (in this case a mixed potential) and $E_2$ would be set to some value greater than $E_1$. The resulting anodic current as a function of time is shown schematically in Fig. 7.9.

The current generally rises instantaneously after the change in potential. The subsequent shape of the current transient will depend on the value of $E_2$ and the nature of the surface on which the oxidation of iron(II) is occurring. Thus, for a reversible (rapid) oxidation reaction at a potential in the diffusion limited region, the current will decay to almost zero as shown by the solid line. More generally the current will decay to some steady-state value depending on the potential.

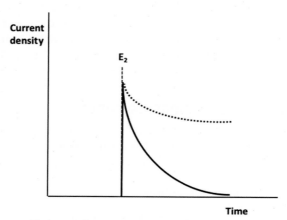

**Figure 7.9** Current responses to potential step.

Once the potential step occurs, iron(II) at the electrode surface is oxidized and a large current begins to flow. However, for the reaction to continue, iron(II) ions need to be replenished at the electrode surface. This happens in stagnant solution via diffusion. It can be shown that the resulting current density-time transient is given by the Cottrell equation

$$i = nF[Fe(II)]_o (D/\pi\, t)^{0.5} \qquad (7.15)$$

in which $[Fe(II)]_o$ is the bulk concentration, D is the diffusion coefficient of iron(II) and t the time.

This relationship that describes semiinfinite linear diffusion show that the current decays with $t^{1/2}$. In the case of a rotating disk electrode (or a stirred solution), the current will not decay to zero because of the applied agitation.

In other cases, the current-time transient is more complex as shown by the example in Fig. 7.10.

In this case, the rapid increase in current at short times is followed by a decrease to low currents at all potentials. While these transients look similar to those predicted by the Cottrell equation, they are not due to a diffusion limited reaction as demonstrated that they are independent of agitation such as the

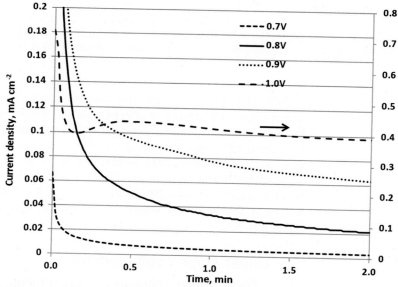

**Figure 7.10** Potentiostatic current-time transients for the oxidation of a pyrite electrode at different potentials in 4M NaCl solution at pH 1. Data: Nicol et al., 2018.

rotation speed of a rotating disk electrode. In this case, slow changes to the composition of the surface of the mineral are responsible for the shapes of the transients.

### 7.3.3 Galvanostatic measurements

In galvanostatic measurements, a fixed current is applied to an electrode and the resulting potential is monitored as a function of time. An example is shown in Fig. 7.11 for the effect of cobalt(II) ions on the rate of evolution of oxygen on a lead alloy electrode such as used in the electrowinning of copper.

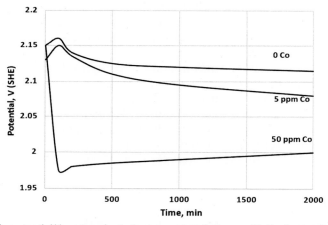

**Figure 7.11** Galvanostatic potential/time transients for oxygen evolution on Pb-Ca-Sn anode in acid solutions containing increasing concentrations of cobalt(II) ions at a constant current density of 280 A m$^{-2}$. Data: Nguyen et al., 2008.

Unlike the previous case, the current due to the evolution of oxygen increases rapidly at short times but attains relatively stable steady-state values at longer times. The presence of cobalt(II) in solution catalyses this reaction (the potential is lower at the same current density) and is actually used in practice during the electrowinning of copper (Chapter 14). Galvanostatic measurements are not that common except in the battery industry for which discharge at constant current is a frequently used evaluation.

### 7.3.4 Linear sweep voltammetry

In a linear sweep voltammetric experiment, the potential applied to the working electrode varies linearly with time from an initial value $E_1$ to a final value $E_2$ as shown in Fig. 7.12.

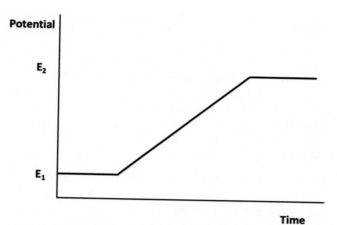

**Figure 7.12** Linear potential sweep experiment.

The sweep rate can vary from as low as 0.1 mV/s to as high as 1 V/s. The current is recorded as a function of time which is proportional to the potential. High sweep rates are to be avoided if possible because of the high double-layer charging currents that can swamp faradaic currents due to the electrode reactions.

An example is shown in Fig. 7.13 that shows the current response (anodic current as positive and cathodic negative) when the potential of a lead alloy electrode initially at either 1.7 V or 2.0 V is swept in a negative direction to a final potential of 1.3 V.

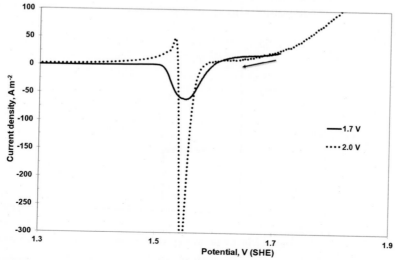

**Figure 7.13** Linear sweep voltammogram of a Pb alloy electrode in acid solution after 10 min potentiostatic at different potentials. The sweeps were initiated in a negative direction at a rate of 10 mV s$^{-1}$. Data: Nguyen et al., 2008.

Peaks can be identified on the voltammogram that vary in position and magnitude depending on the reactions involved. The position of each peak is characteristic of the particular reaction responsible for the current and the magnitude of the peak (or, more precisely, the charge (current × time) under the peak) is proportional to the amount of chemical change. In many cases, it is possible to identify the reactions involved in each peak by comparing the peak potential to the calculated equilibrium potential for the proposed reaction. Thus, for example, the peak at 1.55 V is due to the cathodic reaction

$$PbO_2(s) + SO_4^{2-} + 4H^+ + 2e = PbSO_4(s) + 2H_2O \qquad (7.16)$$

that has a calculated equilibrium potential of about 1.70 V (Table 3.2 in Chapter 3) and the charge under the peak will provide information on the amount of $PbO_2$ present on the surface before reduction. The effects of sweep rate and total charge on the positions and shapes of the peaks often yield useful information on the mechanism of the reaction. In many cases, this assignment of a reaction to a peak is not that obvious and other methods have to be employed to unambiguously make such an assignment. It is important to always specify the starting potential and the direction of the sweep.

For a simple reaction such as

$$Fe(II) = Fe(III) + e$$

A slow linear sweep voltammogram under well-defined mass transport conditions in the absence of iron(III), will give rise to the curves shown in Fig. 7.14.

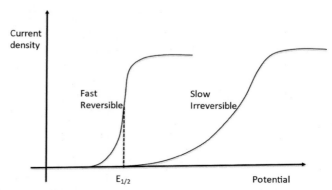

**Figure 7.14** Linear sweep voltammograms under defined mass transport.

Thus, if the rate of electron transfer is rapid (i.e. a high exchange current density), the typical s-shaped curve would be obtained in which the half-wave potential is at the equilibrium potential for the couple-this is often termed a reversible system. For systems with slower kinetics, the wave extends to higher potentials with greater overpotentials required to drive the reaction-an irreversible system. If iron(III) is also present, there will be corresponding cathodic waves in addition to those shown in Fig. 7.14.

For a stationary electrode or at high sweep rates, the characteristic diffusion plateau becomes a peak as shown in Fig. 7.15.

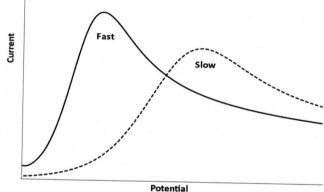

**Figure 7.15** Linear sweep voltammograms at a stationary electrode for fast and slow reactions.

For a highly reversible system, the peak appears at the equilibrium potential. The height of the peak varies with the sweep rate and, for a reversible system, $I_{peak}$ (A) is proportional to the concentration of iron(II) (mol cm$^{-3}$) and also to the square root of the sweep rate ($v$) in V s$^{-1}$.

$$I_p = 2.686 \times 10^5 n^{3/2} A \cdot C \cdot D^{1/2} \, v^{1/2} \qquad (7.17)$$

## 7.3.5 Cyclic voltammetry

In a particular variation of linear sweep voltammetry, the sweep is reversed at $E_2$ and ended at $E_3$. The sweep rates in the forward and reverse direction can be different. A typical waveform is shown in Fig. 7.16.

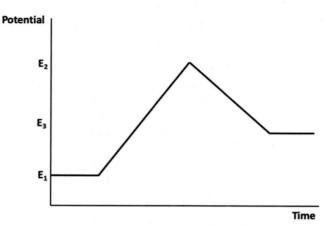

**Figure 7.16** A cyclic voltammetric experiment.

There are a number of variations to this method such as changes to the individual forward and reverse sweep rates, the initial and final potentials, the degree of agitation during each half-cycle, and the number of cycles.

An example is shown in Fig. 7.17 for the deposition of nickel and/or cobalt on a rotating gold disk electrode. In this case, the OCP was −0.3 V and the potential was swept at 1 mV s$^{-1}$ from the OCP in a negative direction to −1.0 V and then back in a positive direction to 0.55 V at the same sweep rate.

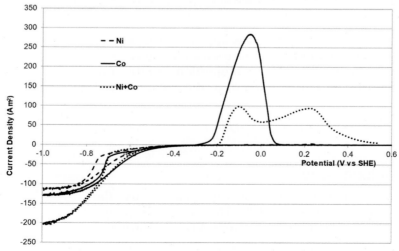

**Figure 7.17** Cyclic voltammogram for the deposition and anodic stripping of nickel and cobalt from dilute sulphate solutions onto a rotating gold disk. The potential was swept at 1 mV s$^{-1}$ from −0.3 V in a negative direction to −1.0 V and then back in a positive direction to 0.55 V. Data: Nicol and Tjandrawan, 2014.

As the potential is swept in a negative direction the cathodic current increases at potentials below about 0.5 V and a plateau is observed at potentials below about −0.8 V. This is due to the mass controlled deposition of nickel and/or cobalt on the gold surface. During the reverse (positive–going) sweep, the metals can be anodically stripped from the gold surface. In the case of cobalt, an anodic peak at about −0.1 V is observed that is due to the dissolution of cobalt deposited during the negative sweep. In the case of nickel, no such peak is observed because nickel is passive under these conditions. When both metals are present, a peak for the stripping of nickel can be seen at about 0.2 V. There are several additional important aspects of these cyclic voltammograms that will be discussed at a later stage.

For a simple reaction such as

$$Fe(II) = Fe(III) + e$$

at a stationary electrode, the current peak for the anodic reaction will be observed as before but, during the negative-going sweep of the cyclic voltammogram, a smaller peak due to the reduction of iron(III) formed during the anodic reaction will also be observed as shown in Fig. 7.18.

For a reversible system at 25°C, the peak separation $\Delta E_{peak} = 0.0591/n$ V where n is the number of electrons involved in the reaction. For irreversible systems, the peak separation will be considerably larger. The size of the cathodic peak will obviously depend on the sweep rate and positive limit of the sweep potential. Thus, at higher sweep rates, less iron(III) will diffuse from the electrode surface before it is reduced during the negative-going sweep.

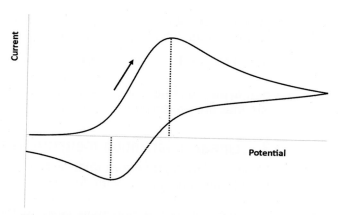

**Figure 7.18** Cyclic voltammogram for a stationary electrode.

### 7.3.6 Coulometry

The objective of coulometric analysis is to establish the stoichiometry of an electrode reaction. This generally involves carrying out either a potentiostatic or galvanostatic experiment under appropriate conditions for a fixed period. The total charge passed is obtained by integration of the area under the current-time transient and the solution analysed for one of the products of reaction. Comparison of the charge with the amount of species analysed then enables the stoichiometry to be established.

For example, Fig. 7.19 shows the stoichiometry of the anodic oxidation of pyrite in terms of the number of Faradays per mol of iron dissolved at various potentials in potentiostatic experiments over several hours.

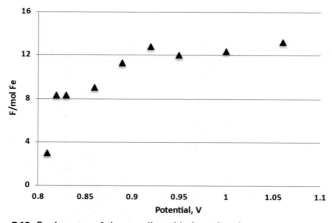

**Figure 7.19** Coulometry of the anodic oxidation of pyrite. Data: Nicol et al., 2018.

These results show that the accepted stoichiometry of 14 e/mol Fe, based on the following anodic reaction

$$FeS_2 + 8H_2O = Fe^{2+} + 2HSO_4^- + 14H^+ + 14e \qquad (7.18)$$

is only achieved at high potentials and that lower oxidation states of sulphur are produced at lower potentials.

### 7.3.7 Linear polarization measurements

A convenient method for rapidly assessing dissolution rates in a mixed potential system is the linear polarization method. This follows from the observation that a combination of two exponential curves is linear at potentials close to the mixed potential.

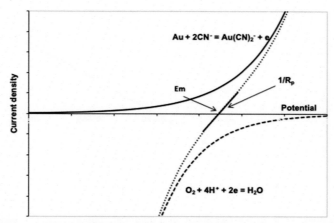

**Figure 7.20** Schematic current—potential curves showing the linear region close to the mixed potential and the polarization resistance.

The reciprocal of the slope of this linear relationship is known as the polarization resistance ($R_p$) and is proportional to the dissolution rate. This is shown in the schematic diagram in Fig. 7.20.

The following relationship can easily be derived for the case in which both half-reactions are in the Tafel region at the mixed potential,

$$i_d = \frac{(b_a\, b_c)}{2.303 R_p\, (b_a + b_c)} \tag{7.19}$$

in which $i_d$ is the dissolution current density (or rate) in A cm$^{-2}$, $b_a$ and $b_c$ are the Tafel slopes for the anodic and cathodic reactions respectively in V/decade and $R_p$ is the polarization resistance in ohm cm$^2$.

An example is shown in Fig. 7.21 for the dissolution of gold in an oxygenated cyanide solution.

The slope of the line can be used to estimate the rate of gold dissolution using Eq. (7.19).

In many cases, it is not necessary to independently measure the Tafel slopes if one can measure the rate of dissolution under a single condition and use this value to calibrate the above equation.

## 7.4 Impedance measurements

This section has been expanded to provide more detailed information given that these measurements are often not very

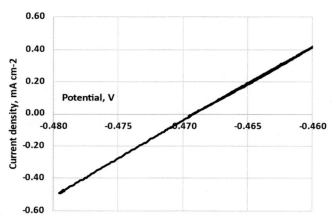

**Figure 7.21** Linear polarization curve for the dissolution of a gold electrode in a cyanide solution saturated with oxygen at pH 10.

well understood by hydrometallurgical researchers. Thus, both the theory behind impedance measurements and the interpretation will be explored in more detail.

Ohm's law gives a simple relation between dc-potential or voltage (E) and dc-current (i):

$$E = i\,R \tag{7.20}$$

where R is the resistance.

When ac-signals are involved the relation is:

$$E_{ac} = i_{ac}\,Z \tag{7.21}$$

where Z is the impedance.

## 7.4.1 Resistor

The most simple case is found with a resistor (R = Z). In this case, the plot of potential for excitation with an AC current (or the reverse) is shown in Fig. 7.22.

Note that in this case, the Phase Angle between I and E, $\varphi = 0$, that is the potential and current are in-phase and the impedance Z of a resistor is independent of frequency. The magnitudes of I and E are simply related by Ohms Law.

In electrochemical cells, the resistance of the solution between the electrodes ($R_s$) or of the electrode itself is the most common source of this impedance.

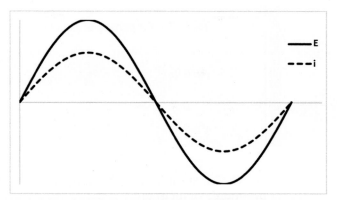

**Figure 7.22** Current–potential curves for the sine wave perturbation of a resistor.

## 7.4.2 Capacitor

The interface between an electrode and solution ideally behaves as a capacitor and is called the electrical double layer capacitance. The magnitude of the impedance of a capacitor is given by

$$Z'' = 1/\omega C \qquad (7.22)$$

in which $\omega$ is the frequency and C the capacitance.

In this case, the phase angle is always 90 degree, that is the potential lags the applied current (or the reverse) as shown in Fig. 7.23.

Note that at high frequencies a capacitor behaves like a resistor with a low value, that is it acts like a short-circuit.

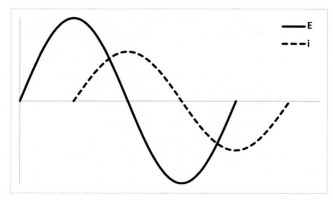

**Figure 7.23** Current–potential curves for the sine wave perturbation of a capacitor.

**Table 7.1 Properties of common circuit components.**

| Component | E vs I | Impedance |
|-----------|--------|-----------|
| Resistor | $E = IR$ | $R$ |
| Capacitor | $I = C\, dE/dt$ | $1/\omega C$ |
| Inductor | $E = L\, dI/dt$ | $\omega L$ |

Table 7.1 summarizes the properties of the most common components of a circuit.

### 7.4.3 Application to electrochemical cells

The simplest electrochemical cell can be simulated by a resistance in series with a capacitor.

$R_s$  $C_{dl}$

in which $R_s$ is the ohmic resistance of the solution and $C_{dl}$ the double layer capacitance.

This corresponds to a completely polarized electrode, that is no steady-state current flows (blocked by the capacitor) when a dc potential difference is imposed.

At high frequencies, the magnitude of the impedance is $R_{sol}$ with a phase angle of 0 degree while at low frequencies, the impedance will be frequency-dependent and the phase angle will be 90 degree. At intermediate frequencies, the phase angle will have a value between 0 and 90 degree.

The complex notation of the impedance of the cell is given by

$$Z = R_{sol} - j/\omega C_{dl} \tag{7.23}$$

Due to the double layer capacitance, the cell impedance usually depends on the applied frequency.

The value of the so-called real impedance, $Z' = R_{sol}$ is independent of the frequency while the quadrature impedance, $Z' = 1/\omega C$ increases with decreasing frequency. These can be plotted as shown in Fig. 7.24. The cell impedance (Z) is given by

$$Z^2 = (Z')^2 + (Z'')^2 \tag{7.24}$$

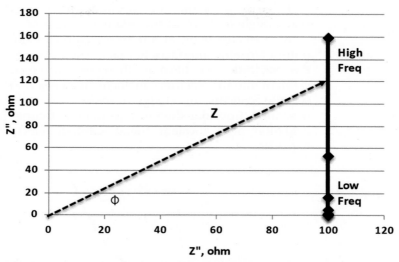

**Figure 7.24** Impedance plot showing the impedance and phase angle of a simple resistor (100 ohm) in series with a capacitor (10 µF). This plot is called Nyquist, Sluyters, Cole-Cole or complex plane plot.

Note that this method of presentation does not explicitly show the effect of frequency. An alternative, often used presentation is shown in Fig. 7.25 as the so-called Bode plot in which the magnitude of the impedance and the phase angle is plotted as a function of the frequency.

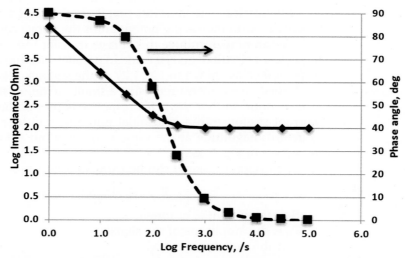

**Figure 7.25** Bode plot for a simple series resistor (100 ohm) and capacitor (10 µF).

It is easy to see that the impedance tends to 100 ohm and the phase angle to 0 deg as the frequency increases.

Real electrochemical anodes or cathodes do not behave as non-polarizable interfaces in that current can be passed through the electrode. We have to therefore introduce an additional element into the electrical analogue of an electrode. Before doing so, the concept of charge transfer resistance needs to be described. Consider the current–potential curve for the anodic oxidation of a divalent metal sulphide electrode shown in Fig. 7.26.

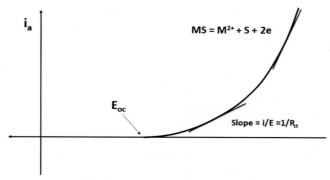

**Figure 7.26** Current/potential curve for an anodic reaction showing the charge transfer resistance.

The slope of the plot at any potential has the units of reciprocal resistance and this resistance is termed the charge transfer resistance ($R_{ct}$). It is the resistance to the flow of current due to the electrochemical reaction. Note that $R_{ct}$ is potential dependent and decreases with increasing potential for the above case.

We can now introduce this charge transfer resistance into the equivalent circuit of an electrode in parallel with the double-layer capacitance as shown in Fig. 7.27 as the so-called Randles cell.

Impedance is measured using a small excitation signal (5 mV). This is done so that the cell's response is pseudo-linear and the current response to a sinusoidal potential will be a sinusoid of different magnitude at the same frequency but shifted in phase as shown in Fig. 7.23.

The excitation signal can be expressed as a function of time,

$$E_t = E_o \sin(\omega t) \tag{7.25}$$

$E_t$ is the potential at time t,
$E_0$ is the amplitude of the signal,
$\omega$ is the radial frequency.

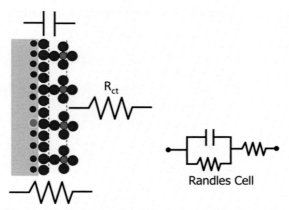

**Figure 7.27** Equivalent circuit for a polarizable electrode.

Radial frequency $\omega$ (expressed in radians $s^{-1}$) is related to the frequency f (expressed in hertz) by

$$\omega = 2 f \qquad (7.26)$$

In a **linear** system, the response signal, $I_t$, is shifted in phase (f) and has a different amplitude, $I_0$.

$$I_t = I_o \sin(\omega_t + f) \qquad (7.27)$$

Using an expression analogous to Ohm's Law, the impedance of the system is

$$Z = \frac{E_t}{I_t} = \frac{E_o \sin(wt)}{I_o \sin(wt + \phi)} = Z_o \frac{\sin(wt)}{\sin(wt + \phi)} \qquad (7.28)$$

The impedance is therefore expressed in terms of a magnitude, $Z_o$, and a phase shift, $\phi$.

With Eulers relationship, it is possible to express the impedance as a complex function. The potential is described as

$$E_t = E_o \exp(j \, \omega \, t) \qquad (7.29)$$

and the current response as,

$$I_t = I_o \exp(j \, \omega \, t - \phi) \qquad (7.30)$$

The impedance is then represented as a complex.number,

$$Z = E/I = Z_o \exp(j \, \phi) = Z_o \cos(\phi) - Z_o \, j \sin(\phi) \qquad (7.31)$$
$$\text{Real(Z')} \quad \text{Imaginary(Z'')}$$

Note the so-called complex impedance notation:

$$Z = ZZ' + jZ'' \text{ with } j = \sqrt{-1} \tag{7.32}$$

The value $Z'$ is the projection along the x-axis and is called in-phase or real impedance.

The value $Z''$ is the projection along the y-axis and is the out-of-phase or quadrature impedance.

The cell impedance is:

$$Z^2 = Z'^2 + Z''^2 \text{ and } \tan \phi = Z''/Z' \tag{7.33}$$

For a Randles cell,

$$Z' = R_{sol} + R_{ct}/\{1 + (\omega R_{ct}C_{dl})^2\} \tag{7.34}$$

$$Z'' = j\omega R_{ct}^2 C_{dl}/\{1 + (\omega R_{ct}C_{dl})^2\} \tag{7.35}$$

The expression for $Z$ is composed of a **real** and an **imaginary** part. If the real part is plotted on the X-axis and the imaginary part is plotted on the Y-axis of a chart, we get a semicircle known as a 'Nyquist Plot' shown in Fig. 7.28. Notice that each point on the Nyquist plot is the impedance at one frequency.

Note that:

At high frequencies,

$$Z^{\wedge\prime} \rightarrow R_{sol} \text{ and } Z^{\wedge\prime\prime} \rightarrow 0$$

At low frequencies,

$$Z^{\wedge\prime} \rightarrow R_{sol} + R_{ct} \text{ and } Z^{\wedge\prime\prime} \rightarrow 0$$

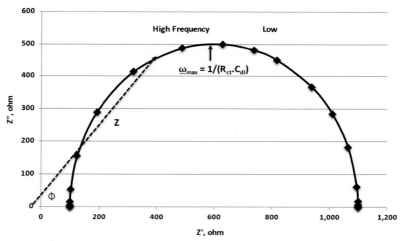

**Figure 7.28** Nyquist plot for a Randles cell with $R_s = 100$ ohm, $R_{ct} = 1000$ ohm and $C_{dl} = 10$ μF.

These two points are the intersections of the semi-circle on the x-axis.

An example of such data is given by the Nyquist plot shown in Fig. 7.29 for the zinc/zinc amalgam electrode in acid solution. This is one of the earliest applications of impedance methods.

In this case, $R_s$ is about 0.85 ohm while $R_s + R_{ct}$ is about 3.3 ohm. Thus, $R_{ct}$ is 2.45 ohm. Values obtained at several concentrations are shown in Table 7.2.

The product of $R_{ct}.[Zn^{2+}]$ is constant confirming that the rate of the reaction is first order in the zinc concentration. The mean value can be converted to a rate constant $k = 3.26 \times 10^{-3}$ cm s$^{-1}$ that is significantly lower than a calculated mass transfer coefficient.

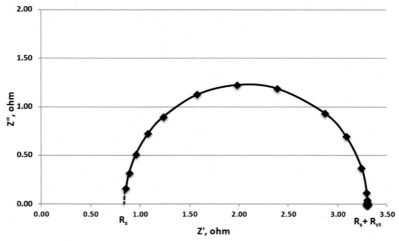

**Figure 7.29** Impedance of a Zn(Hg)/Zn$^{2+}$ couple in 1M NaClO$_4$ at pH 3, [Zn$^{2+}$] = 8mM. Data: Sluyters and Oomen, 1960.

**Table 7.2  Charge transfer resistances for the zinc/zinc amalgam system.**

| [Zn$^{+2}$] (mM) | $R_{ct}$ (ohm cm$^{-2}$) | $R_{ct}.[Zn^{2+}]$ (mM ohm cm$^{-1}$) |
|---|---|---|
| 2 | 10.2 | 20.3 |
| 4 | 4.95 | 19.8 |
| 8 | 2.45 | 19.3 |
| 16 | 1.27 | 20.3 |

## 7.4.4 Effect of mass transport

Diffusion also can create an impedance called a **Warburg** impedance that depends on the frequency of the potential perturbation. At high frequencies, the Warburg impedance is small since diffusing reactants do not have to move very far while at low frequencies, the reactants have to diffuse farther, increasing the Warburg impedance. This impedance is in series with the charge transfer resistance giving the equivalent circuit shown in Fig. 7.30.

The equation for the 'infinite' Warburg impedance is

$$Z_w = \sigma\, \omega^{-1/2} - j\, \sigma\, \omega^{-1/2} \tag{7.36}$$

$\sigma$ is a coefficient that is a function of the concentration (C) and diffusion coefficient (D) of the diffusing species

$$\sigma = RT/(1.416n^2F^2)[1/(C \cdot D^{1/2})]$$
$$= 150 \text{ for C} = 0.1\text{mM and D} = 5 \times 10^{-6} \text{ cm}^2/\text{s} \tag{7.37}$$

On a Nyquist plot, the Warburg impedance appears as a diagonal line with slope of 45° at low frequencies (Fig. 7.31) while it

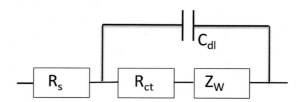

**Figure 7.30** Equivalent circuit including Warburg impedance.

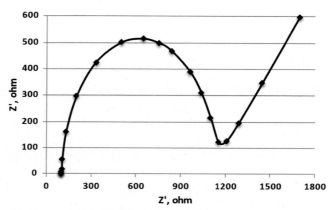

**Figure 7.31** Nyquist plot showing the Warburg impedance. $R_{sol}$ = 100 ohm, $R_{ct}$ = 1000 ohm, $C_{dl}$ = 10 µF, $\sigma$ = 150 ohm s$^{-0.5}$cm$^2$.

exhibits a phase shift of 45 degree on a Bode plot. Note that the plot does not appear to be a semicircle-this is simply due to the different scales on the x- and y-axes.

In cases in which the rate of electron transfer is so fast, the reaction is mass transport controlled at all frequencies and the plot is simply a straight line.

## 7.4.5 Equipment for impedance measurements

Early equipment for impedance measurements generally consisted of the following.
- Potentiostat
- Sine wave generator
- Time synchronization (phase lock amplifier)

Modern instruments often contain the above incorporated into potentiostats and can be considered as 'All-in-ones' or a frequency response analyser with appropriate software for control and analysis of the data. These instruments record very accurate data but the operator must be continually aware of whether the data is meaningful and that it correctly reflects the system under study. Considerable familiarity with electrical systems is necessary in order to get accurate impedance data, particularly at higher frequencies.

## 7.4.6 Network modelling

As has been described above, electrochemical cells can be modelled as a network of passive electrical circuit elements in a network called an 'equivalent circuit' and the technique often known as Electrochemical Impedance Spectroscopy (EIS).

The EIS response of an equivalent circuit can be calculated and compared to the actual EIS response of the electrochemical cell. Complex systems may require complex models in which each element in the equivalent circuit should correspond to some specific activity or component in the electrochemical cell.

It is not acceptable to simply add elements until a good fit is obtained but to use the simplest model that fits the data.

There are dangers in the indiscriminate use of analogues to describe electrochemical systems as equivalent circuits are seldom unique. Only the simplest circuits can be said to be unambiguous in their description of experimental data, and it is possible for several circuits to have the same EIS response.

## 7.4.7 Precautions and limitations of EIS

There are a number of factors that need to be considered in the use of EIS methods that are often overlooked given the ease of operation of modern automated instruments. The most important of these are

**a.** Electrolytes and interfaces are only approximately modelled by idealized circuit elements.

**b.** The effect of geometry (including roughness) on the current distribution can result on frequency dispersion of the impedance. For example, conductivity and interface capacitance should be independent of frequency but often show significant deviations. The introduction of arbitrary CPE (Constant phase elements) is often used to correct for such behaviour.

**c.** Measuring an EIS spectrum takes time (often many hours) and the system must be at a steady state throughout the time required to measure the EIS spectrum. If the cell changes significantly during the time taken to run the impedance test, the results taken at the start of the test (high-frequency part of the sweep) may not be consistent with the results at the end of the test (low frequency).

**d.** A common cause of problems in EIS measurements and analysis is drift in the system. The cell can change through adsorption of solution impurities, growth of a layer, build-up of reaction products in solution, changes in surface morphology.

**e.** Emphasis is placed on identifying an equivalent 'electrical circuit' for the interface. Although the equivalent circuit is capable of mimicking the behaviour of the system, the circuits that often are adopted are too simplistic to be of any interpretive value. For example, in the case of mixed potential systems, a simple Randles circuit is not appropriate because the circuit does not delineate the partial anodic and cathodic reactions involved in a corrosion process. This applies to most oxidative leaching processes as we shall see at a later stage.

## 7.4.8 Capacitance measurements

Measurement of the capacitance of the double-layer can be made at a suitable single frequency depending on the system to provide qualitative information on the properties of the interface. These measurements can be made rapidly such as during a potential sweep (does not require steady state).

The capacitance of an ideal parallel plate capacitor is given by

$$\text{Capacitance} = \varepsilon A/d \qquad 7.39$$

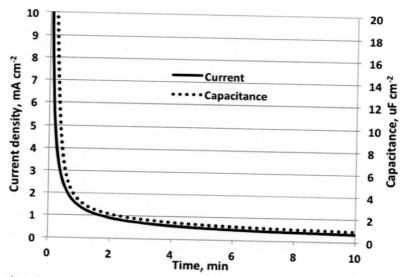

**Figure 7.32** Potentiostatic current transients for oxidation of a rotating chalcopyrite disk electrode. Data: Nicol, 2017.

in which A is the area of each plate, d the distance between the plates and $\varepsilon$ is the dielectric constant of the medium between the plates.

Note that the capacitance decreases with increasing thickness and this is useful information as shown by the data in Fig. 7.32. In this case, the declining current transient during the anodic oxidation of a chalcopyrite electrode is associated with a parallel decrease in the capacitance indicating that this is probably the result of the formation of a pseudo-passive layer that is increasing in thickness with time. The low capacitance (a normal double-layer capacitance would be expected to be in the range 40 to 100 $\mu F\ cm^{-2}$) is also indicative of a solid surface layer.

# 7.5 Practical considerations

The practice of electrochemistry is experimentally subject to many pitfalls and often the data obtained are of little value due to lack of understanding and care in the set-up of experimental work. The following are only a few of the more important considerations.

### 7.5.1 Reference electrodes

Experience has shown that many of the problems encountered in carrying out electrochemical measurements have their origin in the reference electrode. This should be the first step in debugging a setup that is not responding as could be expected. The reference electrode should be periodically checked against a similar electrode maintained as a 'standard'. Deviations of more than 5−10 mV from the quoted reference potential will require a new electrode. Some of the other problems encountered are blocking of the glass or porous plastic fritted junction, air bubbles in the Luggin capillary, loss of the electrolyte in the reference electrode and crystallization of the filling solution in the reference probe resulting in a very high impedance (resistance) that will create unacceptable electrical noise in the measurements.

Measurements at elevated temperatures can create problems in that the reference electrode should be at the same temperature as the electrolyte in the cell or at room temperature if a Luggin capillary is used.

### 7.5.2 Uncompensated resistance

Consider the schematic of a cell in Fig. 7.33 in which a current I is passed through the counter/working electrodes and the potential ($E_{obs}$) of the working electrode is measured with respect to some reference electrode. The resistance of the solution between the tip of the reference electrode and the working electrode is $R_s$. The measured $E_{obs}$ is given by the equation shown in the diagram. It is apparent that the measured potential of the working

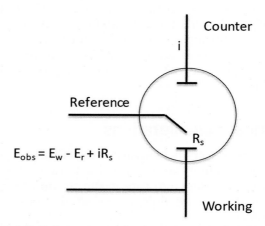

$$E_{obs} = E_w - E_r + iR_s$$

**Figure 7.33** Cell showing origin of uncompensated resistance.

electrode will be greater than the desired value by voltage drop in the solution given by $iR_S$. This uncompensated resistance will obviously be small for low currents and/or high solution conductivity. In many cases, this iR drop must be compensated either manually or automatically by the instrument.

The resistance can be reduced by use of a Luggin capillary as previously mentioned but, at high currents, this is often not sufficient. Automatic compensation is often available on some instruments but can introduce noise and instability in the potentiostat as it relies on positive feedback. The best method appears to be one in which $R_s$ is obtained from the automatic system and then the potential for each data point is manually corrected.

An example of this effect can be seen in the voltammograms shown in Fig. 7.34 for the anodic oxidation of chalcopyrite.

The difference in the slopes of the curves is quite significant with the correction for the iR drop being about 200 mV at 60 A cm$^{-2}$. In this case, part of the uncompensated resistance is due to the resistance of the chalcopyrite electrode itself.

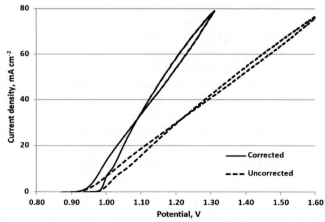

**Figure 7.34** Anodic oxidation of chalcopyrite in 0.1M $H_2SO_4$ at high potentials showing the effect of the uncompensated resistance on the voltammogram. Data: Nicol, 2017.

## 7.5.3 Electrode surface preparation

There are no accepted rules for mineral electrode preparation as each system is different. It is often not necessary to extensively polish the surface with fine polishing powders. On the other hand, simply grinding the surface with coarse grit papers is also not recommended because of the high and non-reproducible surface area generated. A good compromise is to use a medium grit wet paper such as 1200 grit to clean the surface and then 'polish' with 3000 or 4000 grit wet paper. Ultrasonic treatment for a few seconds will remove loosely adhering solids. In many cases, the availability of electrode materials is limited and one has to conserve the electrode for use in multiple experiments. The exposed geometric surface area can be obtained by simple measurement if regular or by photographing the surface and deriving the area from the photograph by suitable means if irregular.

## 7.6 Summary

This Chapter has been included because of the widespread use of electrochemical methods to study the many reactions in hydrometallurgy that involve the transfer of charge across interfaces. The reader should now be familiar with the nature of electrochemical experiments including the types of cells, 3-electrode systems and reference electrodes. The principle of operation of potentiostats and galvanostats has been outlined and the more common electrochemical experiments including equilibrium potential measurements, potentiostatic and galvanostatic techniques including coulometry, linear sweep and cyclic voltammetry have been described. The basis of impedance measurements and their advantages and disadvantages have been covered that should enable the reader to more easily assess the relevance of such measurements in the study of the complex electrochemical reactions that are characteristic of hydrometallurgical systems. Of particular importance is the application of electrochemistry to oxidative/reductive dissolution of conducting solids and the mixed potential model as applied to leaching, cementation, gaseous and ionic reduction.

## References

Nicol, M.J., Tjandrawan, V., 2014. The effects of thiosulfate ions on the deposition of cobalt and nickel from sulfate solutions. Hydrometallurgy 150, 34–40.

Nicol, M.J., 2017. The anodic behaviour of chalcopyrite in chloride solutions: overall features and comparison with sulfate solutions. Hydrometallurgy 169, 321–329.

Nicol, M.J., Zhang, S., Tjandrawan, V., 2018. The electrochemistry of pyrite in chloride solutions. Hydrometallurgy 178, 116–123.

Nguyen, T., Guresin, N., Nicol, M., Atrens, A., 2008. Influence of cobalt ions on the anodic oxidation of a lead alloy under conditions typical of copper electrowinning. J. Appl. Electrochem. 38, 215–224.

Sluyters, J.H., Oomen, J.J.C., 1960. On the impedance of galvanic cells. II. Experimental verification. Rec. Trav. Chim 79, 1101–1110.

## Further reading

Gabrielli, C. Electrochemical Impedance Spectroscopy: Principles, Instrumentation and Applications.

Macdonald, D.D., 1977. Transient Techniques in Electrochemistry. Plenum Press, New York.

Oldham, K.B., Myland, J.C., 1994. Fundamentals of Electrochemical Science. Academic Press, San Diego.

## Case study

Fig. 7.35 shows schematically the current–potential curves for the dissolution of pyrite in acidic solutions using iron(III) and dissolved oxygen as the oxidant.

**(a)** Write balanced equations for the overall reactions for both iron(III) and oxygen.
**(b)** Which oxidant is more effective? Why?
**(c)** Estimate the initial rate of dissolution of pyrite particles (50 μm diameter) under these conditions.

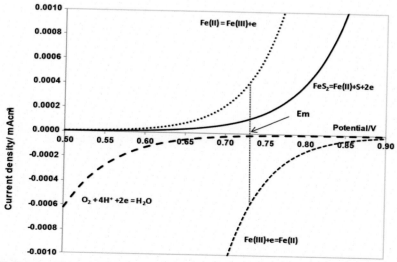

**Figure 7.35** Schematic current–potential curves for the dissolution of pyrite by iron(III) and oxygen.

(d) Assuming that the rate is not affected by the sulphur layer on the surface, estimate the time for complete dissolution of a particle.

(e) The following data was obtained from a slow sweep voltammogram for the cathodic reduction of dissolved $O_2$ ($0.9 \times 10^{-3}$M) at a rotating pyrite disk at 25°C in a dilute NaCl solution at pH 1.

| Potential (V vs SHE) | Current density (A cm$^{-2}$) |
| --- | --- |
| 0.55 | −0.025 |
| 0.50 | −0.030 |
| 0.45 | −0.055 |
| 0.40 | −0.105 |
| 0.35 | −0.210 |
| 0.30 | −0.420 |
| 0.25 | −0.750 |

Calculate the exchange current density for the reduction of oxygen under these conditions. Estimate the current density for the reduction of oxygen at the mixed potential shown in Fig. 7.33. How does this compare with the overall current density at $E_m$?

You should note any assumptions made in the above calculations.

# Index